AF306380

New Euclidon Method of Generating Stationary Vacuum Einstein Fields

New Euclidon Method of Generating Stationary Vacuum Einstein Fields

Tsarai I Gutsunaev
Peoples' Friendship University of Russia, Russia

Aleksandr A Shaideman
Peoples' Friendship University of Russia, Russia

Kirill V Golubnichiy
Texas Tech University, USA

World Scientific

NEW JERSEY · LONDON · SINGAPORE · BEIJING · SHANGHAI · HONG KONG · TAIPEI · CHENNAI · TOKYO

Published by

World Scientific Publishing Co. Pte. Ltd.

5 Toh Tuck Link, Singapore 596224

USA office: 27 Warren Street, Suite 401-402, Hackensack, NJ 07601

UK office: 57 Shelton Street, Covent Garden, London WC2H 9HE

Library of Congress Control Number: 2025942023

British Library Cataloguing-in-Publication Data
A catalogue record for this book is available from the British Library.

NEW EUCLIDON METHOD OF GENERATING STATIONARY VACUUM EINSTEIN FIELDS

ISBN 978-981-98-1391-9 (hardcover)
ISBN 978-981-98-1392-6 (ebook for institutions)
ISBN 978-981-98-1393-3 (ebook for individuals)

For any available supplementary material, please visit
https://www.worldscientific.com/worldscibooks/10.1142/14336#t=suppl

Desk Editor: Muhammad Ihsan Putra

Typeset by Stallion Press
Email: enquiries@stallionpress.com

"In memory of the scientists Ya. P. Terletsky and M. F. Sukhinin and the teacher M. A. Konopleva, the author of the Handbook on Modern School Mathematics."

Preface

A fault-finding reader will hardly find in this book a systematic account of the foundations of general theory of relativity. The proposed book is not also a review-article on exact solutions of the general relativity equations ordered by some physical journal. It is neither a textbook nor a manual. Most likely it represents the complex of results of our investigations carried out recently in the sphere of the Einstein–Maxwell equations.

At present, the general theory of relativity, from our point of view, consists of two parts: the mathematical apparatus and the physical interpretation of its results. What the first part is concerned, it does undoubtedly reflect the objective reality. It is the physical interpretation of the general relativity which is subjected to certain criticism (especially during the last time). That is why we try in our book not to impose on the obtained results a determined physical meaning based on the use of the traditional set of concepts and terms.

When one well-known theoretical physicist was asked what was the physical meaning of the Maxwell equations, he answered — the equations themselves. The scientist's reply does not lack a common sense though it looks like a pun. The farther one moves away from the habitual physical reality, the more unexpected sense acquire our orthodox conceptions.

We tried to use the most effective and at the same time the simplest (with respect to calculations) method for obtaining of either already known solutions or the new ones. The primary "building block" used is the Euclidon solution, which has a clear physical interpretation as a relativistic accelerated non-inertial reference frame in the flat spacetime. These solutions serve as "building blocks" of the theory, which allows for the construction of almost all known solutions to the vacuum static axially symmetric Einstein equations, including such important ones as the Schwarzschild and Kerr solutions, and provides a different perspective on the physical interpretation of well-known solutions, such as the Schwarzschild and Kerr solutions. For all that, the mathematical contents of the book prevail over the physical results.

We think that our book will attract attention of our colleagues-relativists working on similar problems. We also hope that it will excite curiosity of all those who are just starting to discover the mysteries of the greatest physical theory and who, having obtained their first even probably insignificant results, have become its captives forever.

Ts. I. Gutsunaev, A. A. Shaideman,
and K.V. Golubnichiy
Moscow
December 2024

Contents

Chapter 1

Static Solutions of the Vacuum Einstein's Equations

1.1 Introduction

The formulation of the general theory of relativity in 1915 by Einstein [1] became a starting point for research in the field of exact solutions of the new theory.

The vacuum static Einstein equations for the case of spherical symmetry were considered in 1916 by Schwarzschild [2] who obtained a solution which turned out to be appropriate for describing the gravitational field around a spherically symmetric mass distribution.

Further research in the field of exact solutions of the general relativity equations was appreciably influenced by the work of Weyl [3]. The static axially symmetric vacuum Weyl equations formed a basis for obtaining new solutions by Chazy [4] and Curzon [5], Erez and Rosen [6] and others (see Refs. [7–10] and the monograph [11]).

1.2 Basic Equations

For axially symmetric vacuum static gravitational fields, the line element reduces to the Weyl metric

$$ds^2 = f^{-1}[e^{2\gamma}(d\rho^2 + dz^2) + \rho^2 d\varphi^2] - f dt^2, \qquad (1.1)$$

where $\rho\varphi, z$ and t are canonical Weyl coordinates and time, respectively, and $f(\rho, z)$ and $\gamma(\rho, z)$ are unknown functions to be determined from the field equations.

The covariant metric tensor g_{ik} in this case has only four non-zero components

$$g_{11} = g_{22} = f^{-1}e^{2\gamma}, \quad g_{33} = f^{-1}\rho^2, \quad g_{44} = -f, \qquad (1.2)$$

and the determinant of matrix $\|g_{ik}\|$ is equal to $g = \det\|g_{ik}\| = -(f^{-1}\rho e^{2\gamma})^2$.

The field equations take the form

$$G_{ik} = 0, \qquad (1.3)$$

where G_{ik} is the Einstein tensor related to Ricci tensor R_{ik} and the curvature scalar R by

$$G_{ik} = R_{ik} - (1/2)g_{ik}R. \qquad (1.4)$$

To calculate the Ricci tensor and curvature scalar, one should use the formulae

$$R_{ik} = \frac{\partial \Gamma^l_{ik}}{\partial x^l} - \frac{\partial \Gamma^l_{il}}{\partial x^k} + \Gamma^l_{ik}\Gamma^m_{lm} - \Gamma^m_{il}\Gamma^l_{km}, \qquad (1.5)$$

$$R = R_{ik}g^{ik}, \qquad (1.6)$$

and the Christoffel symbols Γ^i_{kl} could be found from the relations

$$\Gamma^i_{kl} = \frac{1}{2}g^{im}\left(\frac{\partial g_{mk}}{\partial x^l} + \frac{\partial g_{ml}}{\partial x^k} - \frac{\partial g_{kl}}{\partial x^m}\right). \qquad (1.7)$$

Thus, we obtain the Einstein equations for an axially symmetric gravitational field outside the sources:

$$f\Delta f = (\vec{\nabla}f)^2, \qquad (1.8)$$

$$4\frac{\partial \gamma}{\partial \rho} = \rho f^{-2}\left[\left(\frac{\partial f}{\partial \rho}\right)^2 - \left(\frac{\partial f}{\partial z}\right)^2\right], \quad 2\frac{\partial \gamma}{\partial z} = \rho f^{-2}\frac{\partial f}{\partial \rho}\frac{\partial f}{\partial z}. \qquad (1.9)$$

The operators $\vec{\nabla}$ and Δ are defined by the formulae

$$\vec{\nabla} \equiv \vec{\rho}_0\frac{\partial}{\partial \rho} + \vec{z}_0\frac{\partial}{\partial z},$$

$$\Delta \equiv \vec{\nabla}^2 \equiv \frac{\partial^2}{\partial \rho^2} + \frac{1}{\rho}\frac{\partial}{\partial \rho} + \frac{\partial^2}{\partial z^2}$$

($\vec{\rho}_0$ and $\vec{z}_0$ being unit vectors), i.e., they are similar to the ordinary Laplacian and gradient operators for flat space expressed in cylindrical coordinates provided that there is no angular coordinate dependence.

Note also that the integrability condition of equations (1.9) for determining γ is equation (1.8) which does not contain γ.

Solutions of equations (1.8) and (1.9) define the Weyl class.

With the substitution

$$f = e^{2\psi}, \tag{1.10}$$

equation (1.8) becomes linear:

$$\Delta\psi \equiv \frac{\partial^2\psi}{\partial\rho^2} + \frac{1}{\rho}\frac{\partial\psi}{\partial\rho} + \frac{\partial^2\psi}{\partial z^2} = 0. \tag{1.11}$$

The system (1.9) can be rewritten in the form

$$\frac{\partial\gamma}{\partial\rho} = \rho\left[\left(\frac{\partial\psi}{\partial\rho}\right)^2 - \left(\frac{\partial\psi}{\partial z}\right)^2\right], \quad \frac{\partial\gamma}{\partial z} = 2\rho\frac{\partial\psi}{\partial\rho}\frac{\partial\psi}{\partial z}. \tag{1.12}$$

In 1959, Erez and Rosen [6] presented a procedure for obtaining asymptotically flat solutions of equation, for which they used the prolate ellipsoidal coordinates (x, y).

These coordinates are related to the Weyl canonical coordinates (ρ, z) by the formulae

$$
\begin{aligned}
x &= \frac{1}{2k_0}\left[\sqrt{\rho^2 + (z + k_0)^2} + \sqrt{\rho^2 + (z - k_0)^2}\right], \\
y &= \frac{1}{2k_0}\left[\sqrt{\rho^2 + (z + k_0)^2} - \sqrt{\rho^2 + (z - k_0)^2}\right],
\end{aligned}
\tag{1.13}
$$

where k_0 is a real constant.

The inverse transformation is

$$\rho = k_0\sqrt{(x^2 - 1)(1 - y^2)}, \quad z = k_0 xy. \tag{1.14}$$

In the new coordinates, the operators Δ and $\vec{\nabla}$ take the form

$$\Delta \equiv \frac{1}{k_0^2(x^2 - y^2)}\left(\frac{\partial}{\partial x}\left[(x^2 - 1)\frac{\partial}{\partial x}\right] + \frac{\partial}{\partial y}\left[(1 - y^2)\frac{\partial}{\partial y}\right]\right),$$

$$\vec{\nabla} \equiv \frac{1}{k_0(x^2 - y^2)^{1/2}}\left(\vec{x}_0\sqrt{x^2 - 1}\frac{\partial}{\partial x} + \vec{y}_0\sqrt{1 - y^2}\frac{\partial}{\partial y}\right),$$

$\vec{x}_0$ and $\vec{y}_0$ being unit vectors.

The line element in this case can be rewritten as

$$ds^2 = k_0^2 f^{-1} \left[e^{2\gamma}(x^2 - y^2) \left(\frac{dx^2}{x^2 - 1} + \frac{dy^2}{1 - y^2} \right) \right.$$

$$\left. + (x^2 - 1)(1 - y^2) d\varphi^2 \right] - f dt^2, \tag{1.15}$$

where the metric coefficients f and γ are functions of x and y.

Equations (1.8) and (1.9) become

$$\frac{\partial}{\partial x}\left[(x^2 - 1)\frac{\partial \psi}{\partial x}\right] + \frac{\partial}{\partial y}\left[(1 - y^2)\frac{\partial \psi}{\partial y}\right] = 0, \tag{1.16}$$

$$\frac{\partial \gamma}{\partial x} = \frac{1 - y^2}{x^2 - y^2}\left[x(x^2 - 1)\left(\frac{\partial \psi}{\partial x}\right)^2 - x(1 - y^2)\left(\frac{\partial \psi}{\partial y}\right)^2 \right.$$

$$\left. - 2y(x^2 - 1)\frac{\partial \psi}{\partial x}\frac{\partial \psi}{\partial y} \right], \tag{1.17}$$

$$\frac{\partial \gamma}{\partial y} = \frac{x^2 - 1}{x^2 - y^2}\left[y(x^2 - 1)\left(\frac{\partial \psi}{\partial x}\right)^2 - y(1 - y^2)\left(\frac{\partial \psi}{\partial y}\right)^2 \right.$$

$$\left. + 2x(1 - y^2)\frac{\partial \psi}{\partial x}\frac{\partial \psi}{\partial y} \right]. \tag{1.18}$$

For the physical interpretation of the results of the gravitational equation solutions, the most convenient expressions are those obtained for the field potentials, written in curvature coordinates r, θ, φ, which are related to the stretched ellipsoidal coordinates x, y, φ by the following relations:

$$x = \frac{r}{k_0} - 1, \quad y = \cos \theta, \quad \varphi = \varphi. \tag{1.19}$$

In these coordinates, the operators take the following form:

$$\Delta \equiv \frac{1}{(r^2 - 2k_0 r + k_0^2 \sin^2 \theta)\sin\theta}\left\{ \sin\theta \frac{\partial}{\partial r}\left[(r^2 - 2k_0 r)\frac{\partial}{\partial r}\right] \right.$$

$$\left. + \frac{\partial}{\partial \theta}\left[\sin\theta \frac{\partial}{\partial \theta}\right] \right\},$$

$$\vec{\nabla} \equiv \frac{1}{(r^2 - 2k_0 r + k_0^2 \sin^2 \theta)^{1/2}} \left[\vec{r_0}(r^2 - 2k_0 r)^{1/2} \frac{\partial}{\partial r} + \vec{\theta_0} \frac{\partial}{\partial \theta} \right],$$

where $\vec{r_0}$ and $\vec{\theta_0}$ are unit vectors along the r and θ axes and the Weyl metric is written as follows:

$$ds^2 = f^{-1} \left[e^{2\gamma}(r^2 - 2k_0 r + k_0^2 \sin^2 \theta) \left(\frac{dr^2}{r^2 - 2k_0 r} + d\theta^2 \right) \right.$$

$$\left. + (r^2 - 2k_0 r) \sin^2 \theta d\varphi^2 \right] - f dt^2. \tag{1.20}$$

Equation (1.8), taking into account the given factors, is reduced to the following form:

$$\frac{\partial}{\partial r} \left[(r^2 - 2k_0 r) \frac{\partial \psi}{\partial r} \right] + \frac{1}{\sin \theta} \frac{\partial}{\partial \theta} \left[\sin \theta \frac{\partial \psi}{\partial \theta} \right] = 0. \tag{1.21}$$

To conclude, let us consider the Kretschmann scalar

$$\alpha = R_{iklm} R^{iklm}, \tag{1.22}$$

where $R_{iklm} = g_{in} R^n_{klm}$ and the Riemann tensor R^i_{klm} is related to Christoffel symbols as follows:

$$R^i_{klm} = \frac{\partial \Gamma^i_{km}}{\partial x^l} - \frac{\partial \Gamma^i_{kl}}{\partial x^m} + \Gamma^i_{nl} \Gamma^n_{km} - \Gamma^i_{nm} \Gamma^n_{kl}.$$

The Kretschmann scalar has the form [12]

$$\alpha = 16 e^{-4(\gamma - \psi)} [(R^1_{212})^2 + (R^3_{131})^2 + (R^3_{132})^2 + R^1_{212} R^3_{131}], \tag{1.23}$$

where

$$R^1_{212} = \left(\frac{\partial \psi}{\partial \rho} \right)^2 + \left(\frac{\partial \psi}{\partial z} \right)^2 - \frac{1}{\rho} \frac{\partial \psi}{\partial \rho},$$

$$R^3_{131} = -\frac{\partial^2 \psi}{\partial z^2} + \left(\frac{\partial \psi}{\partial \rho}\right)^2 - 2\left(\frac{\partial \psi}{\partial z}\right)^2 - \rho\frac{\partial \psi}{\partial \rho}\left[\left(\frac{\partial \psi}{\partial \rho}\right)^2 - 3\left(\frac{\partial \psi}{\partial z}\right)^2\right],$$

$$R^3_{132} = \frac{\partial^2 \psi}{\partial \rho \partial z} + 3\frac{\partial \psi}{\partial \rho}\frac{\partial \psi}{\partial z} + \rho\frac{\partial \psi}{\partial z}\left[\left(\frac{\partial \psi}{\partial z}\right)^2 - 3\left(\frac{\partial \psi}{\partial \rho}\right)^2\right].$$

This scalar provides information on internal singularities or gravitational field and depends in the general case on the direction of the approach to the singular point. A detailed physical interpretation of singularities can be found in Refs. [13–16].

1.3 Method of Separation of Variables

The equation

$$\frac{\partial^2 \psi}{\partial \rho^2} + \frac{1}{\rho}\frac{\partial \psi}{\partial \rho} + \frac{\partial^2 \psi}{\partial z^2} = 0 \tag{1.24}$$

can be solved by separation variables.

In this case, we have

$$\psi(\rho, z) = \int_0^\infty c_1(\lambda)I_0(\lambda\rho)\sin(\lambda(z + \alpha_0) + \beta_0)\,d\lambda$$

$$+ \int_0^\infty c_2(\lambda)K_0(\lambda\rho)\sin(\lambda(z + \alpha_0) + \beta_0)\,d\lambda, \tag{1.25}$$

where $I_0(\lambda\rho)$ is the Enfield function and $K_0(\lambda\rho)$ is the McDonald function.

By a particular choice of $c_1(\lambda), c_2(\lambda), \alpha_0, \beta_0$ in (1.25), we obtain all known gravistatic solutions.

Example 1. If $c_1(\lambda) = 0,\ \ c_2(\lambda) = -(2m/\pi),\ \ \alpha_0 = 0$, and $\beta_0 = \pi/2$, then

$$\psi(\rho, z) = -\frac{2m}{\pi}\int_0^\infty K_0(\lambda\rho)\cos(\lambda z)\,d\lambda = -\frac{m}{\sqrt{\rho^2 + z^2}}. \tag{1.26}$$

According to (1.12), the metric function γ in this case is

$$\gamma(\rho, z) = -\frac{m^2 \rho^2}{2(\rho^2 + z^2)^2} \tag{1.27}$$

and the line element (1.1) takes the form

$$ds^2 = \exp\left(\frac{2m}{\sqrt{\rho^2 + z^2}}\right)\left[(d\rho^2 + dz^2)\exp\left(-\frac{m^2 \rho^2}{(\rho^2 + z^2)^2}\right) + \rho^2 d\varphi^2\right]$$
$$- dt^2 \exp\left(-\frac{2m}{\sqrt{\rho^2 + z^2}}\right), \tag{1.28}$$

which is the Chazy–Curzon solution [4, 5]. Its physical interpretation is consider in Refs. [17–22].

The Kretschmann scalar, calculated according to (1.23), is

$$\alpha = \exp\left(\frac{2m}{\sqrt{\rho^2 + z^2}}\left[\frac{m\rho^2}{(\rho^2 + z^2)^{3/2}} - 2\right]\right) \times A, \tag{1.29}$$

where A is the polynomial in $\frac{1}{\sqrt{\rho^2 + z^2}}$.

The physical interpretation of singularities and the behavior of α are considered in Refs. [23, 24].

Example 2. If we choose $\psi(\rho, z)$ in the form

$$\psi(\rho, z) = -m \int_0^\infty \frac{\sin(\lambda i \beta)}{\lambda i \beta} J_0(\lambda \rho) \exp(-\lambda z)\, d\lambda,$$

where $J_0(\lambda \rho)$ is the Bessel function, we get a new solution:

$$\psi(\rho, z) = -\frac{m}{\beta} \ln[u + \sqrt{1 + u^2}], \tag{1.30}$$

where

$$u = \frac{2\beta}{\sqrt{(\rho + i\beta)^2 + z^2} + \sqrt{(\rho - i\beta)^2 + z^2}}, \quad \beta = \text{const.}$$

Now, if we put $\beta = 0$, then (1.30) reduces to the Chazy–Curzon solution.

Example 3. If we choose $c_1(\lambda) = 0$, $c_2(\lambda) = \frac{1}{\pi\lambda}$, and $\alpha_0 = w = \text{const}$, $\beta_0 = 0$, the solution (1.25) takes the form

$$
\psi(\rho, z) = \frac{1}{\pi} \int_0^\infty \frac{\sin \lambda(z + w)}{\lambda} K_0(\lambda\rho)\, d\lambda
$$

$$
= \frac{1}{2} \ln[z + w + \sqrt{\rho^2 + (z + w)^2}] - \frac{1}{2} \ln \rho. \qquad (1.31)
$$

This solution is usually termed a "static soliton" (e.g., Ref. [57]). Since $(1/2)\ln \rho$ solves equation (1.25), the function

$$
\psi(\rho, z) = \frac{1}{2} \ln[z + w + \sqrt{\rho^2 + (z + w)^2}] \qquad (1.32)
$$

also satisfies the above equation. The corresponding function γ reduces to the form

$$
\gamma(\rho, z) = \frac{1}{2} \ln \frac{z + w + \sqrt{\rho^2 + (z + w)^2}}{\sqrt{\rho^2 + (z + w)^2}}. \qquad (1.33)
$$

The corresponding linear element is

$$
ds^2 = \frac{d\rho^2 + dz^2}{\sqrt{\rho^2 + (z + w)^2}} + \frac{\rho^2 d\varphi^2}{z + w + \sqrt{\rho^2 + (z + w)^2}}
$$

$$
- [z + w + \sqrt{\rho^2 + (z + w)^2}] dt^2. \qquad (1.34)
$$

It makes sense to call the solutions (1.32) and (1.33) "static Euclidon" since direct calculation shows that for metric (1.34), all components or Riemann–Christoffel curvature tensor turn to zero. It follows that the spacetime is flat, which makes it possible to obtain transition formulae from the Minkowski interval to metric (1.34), which characterizes some relativistic non-inertial frame.

It is worth noting that the solution $\psi = (1/2)\ln \rho$ also corresponds to the Euclidean space (e.g., Ref. [25]).

It is well known that if a complex function ψ satisfies equation (1.24), then its real and imaginary parts are also solutions of this equation. In Refs. [26, 27], this property was applied for obtaining new solutions.

Thus, one may set the constants of translation w in (1.31) and (1.32) complex, as was done in Ref. [28].

Example 4. The Schwarzschild solution [2] can be obtained from (1.25) by setting $c_1(\lambda) = 0$, $c_2(\lambda) = -\frac{2}{\pi}\frac{\sin(\lambda m)}{\lambda}$, $\alpha_0 = 0$, $\beta_0 = \frac{\pi}{2}$ so that

$$\psi(\rho, z) = -\frac{2}{\pi}\int_0^\infty \frac{\cos(\lambda z)\sin(\lambda m)}{\lambda}K_0(\lambda\rho)\,d\lambda$$

$$= \frac{1}{2}\ln\left[\frac{z - m + \sqrt{\rho^2 + (z - m)^2}}{z + m + \sqrt{\rho^2 + (z + m)^2}}\right]. \tag{1.35}$$

For the metric function γ, one has

$$\gamma(\rho, z) = \frac{1}{2}\ln\left[\frac{(R_+ + R_-)^2 - 4m^2}{4R_+ R_-}\right], \tag{1.36}$$

where

$$R_\pm = \sqrt{\rho^2 + (z \pm m)^2)}.$$

In the coordinates (r, θ) $(x = (r/k_0) - 1$, $y = \cos\theta$, $k_0 = m)$, it takes the form

$$ds^2 = \frac{dr^2}{1 - 2m/r} + r^2(d\theta^2 + \sin^2\theta d\varphi^2) - \left(1 - \frac{2m}{r}\right)dt^2. \tag{1.37}$$

A detailed physical interpretation of solutions (1.35), (1.36) can be found in Refs. [29–32].

The Kretschmann scalar for the Schwarzschild solution is

$$\alpha = \frac{48m^2}{r^6},$$

which corresponds to a collapsed source or a black hole.

The significance of the Schwarzschild metric lies in the fact that according to the Birkhoff [33] theorem, it is the unique static spherically symmetric vacuum solution of the Einstein equations.

Moreover, Israel [34] has shown that no other static vacuum solution has a completely regular horizon.

Example 5. With the help of (1.25), another non-trivial solution can be obtained.

Let us assume $c_1(\lambda) = \frac{e^{-i\lambda\varepsilon_0}}{2i\lambda}$, $c_2(\lambda) = 0$, $\varepsilon_0 = \text{const}$, $\alpha_0 = \pm w$, $\beta_0 = 0$.

Then,

$$\psi_\pm(\rho, z) = \int_0^\infty \frac{\sin\lambda(z \pm w)}{2i\lambda} e^{-i\lambda\varepsilon_0} I_0(\lambda\rho)\, d\lambda$$

$$= \frac{1}{2}\ln\left[\zeta_\pm + \sqrt{\zeta_\pm^2 + \rho^2}\right] - \frac{1}{2}\ln\rho. \qquad (1.38)$$

Note that if we take a combination of two solutions (1.38), we obtain

$$\psi = \frac{1}{2}\ln\left[\frac{\zeta_- + \sqrt{\zeta_-^2 + \rho^2}}{\zeta_+ + \sqrt{\zeta_+^2 + \rho^2}}\right],$$

where

$$\zeta_\pm^2 = \frac{1}{2}\left[\sqrt{(\varepsilon_0^2 + \rho^2 - (w \pm z)^2)^2 + 4\rho^2(w \pm z)^2}\right.$$

$$\left. - (\varepsilon_0^2 + \rho^2 - (w \pm z)^2)\right].$$

If we put $\varepsilon_0 = 0$, $w = m$, then ψ reduces to the Schwarzschild solution (1.32).

Example 6. Equation (1.24) also admits separation of variables in the spherical coordinates (R, ϑ):

$$R = \sqrt{\rho^2 + z^2}, \quad \cos\vartheta = \frac{z}{\sqrt{\rho^2 + z^2}}.$$

Equation (1.24) in the spherical coordinates takes the form

$$\sin\vartheta\frac{\partial}{\partial R}\left(R^2\frac{\partial\psi}{\partial R}\right) + \frac{\partial}{\partial\vartheta}\left(\sin\vartheta\frac{\partial\psi}{\partial\vartheta}\right) = 0.$$

In this case, we have the solution

$$\psi(R, \vartheta) = \sum_{n=0}^\infty [a_0 P_n(\cos\vartheta) + b_0 Q_n(\cos\vartheta)][p_0 R^n + q_0 R^{-n-1}], \quad (1.39)$$

$n = 0, 1, 2, \ldots,$

where a_0, b_0, p_0, q_0 are constants, $P_n(\cos\vartheta)$ are Legendre polynomials, and $Q_n(\cos\vartheta)$ are Legendre functions of second type.

Asymptotically flat solutions in this case are determined by the potential [11]

$$\psi(R,\vartheta) = \sum_{n=0}^{\infty} a_n R^{-n-1} P_n(\cos\vartheta). \tag{1.40}$$

Then, the metric function γ is

$$\gamma(R,\vartheta) = -\sum_{n=0}^{\infty}\sum_{m=0}^{\infty}\left[\frac{a_n a_m(n+1)(m+1)}{(n+m+2)R^{n+m+2}}\right]$$
$$\times\left[P_n(\cos\vartheta)P_m(\cos\vartheta) - P_{n+1}(\cos\vartheta)P_{m+1}(\cos\vartheta)\right]. \tag{1.41}$$

Let us take some particular cases of this soliton:

(a) If $a_0 = -m$, $a_n = 0$, when $n \geq 1$, then

$$\psi = -\frac{m}{R}, \qquad \gamma = -\frac{m^2\sin^2\vartheta}{2R^2}. \tag{1.42}$$

It is the Chazy–Curzon solution (1.26), (1.27) in the spherical coordinates.

(b) If $a_n = -\frac{\delta_0}{2}(n+1)^{-1}m^{n+1}[1+(-1)^n]$, then

$$\psi(\rho,z) = \frac{\delta_0}{2}\ln\left[\frac{R_+ + R_- - 2m}{R_+ + R_- + 2m}\right],$$

$$\gamma(\rho,z) = \frac{\delta_0^2}{2}\ln\left[\frac{(R_+ + R_-)^2 - 4m^2}{4R_+ R_-}\right], \tag{1.43}$$

$$R_\pm = \sqrt{\rho^2 + (z\pm m)^2}.$$

It is the Zipoy solution.

In the case $\delta_0 = 1$, (1.43) tends to the Schwarzschild solution.

Gravistatic solutions in the spherical coordinates (R,ϑ) are discussed in Ref. [37].

Example 7. Consider separation of variables in equation (1.16). In this case, the solution $\psi(x,y)$ can be written as follows:

$$\psi(x,y) = \sum_{l=0}^{\infty}[q_l Q_l(x) + p_l P_l(x)][b_l Q_l(y) + d_l P_l(y)], \tag{1.44}$$
$$l = 0,1,2,\dots,$$

where q_l, p_l, b_l, d_l are constants, $P_l(x)$ are Legendre polynomials, and $Q_l(x)$ are Legendre functions of second type.

Solutions satisfying the asymptotic flatness condition

$$\lim_{x\to\infty} \psi(x,y) = \lim_{x\to\infty} \gamma(x,y) = 0 \tag{1.45}$$

can be written in the form

$$\psi(x,y) = \sum_{l=0}^{\infty} q_l Q_l(x) P_l(y). \tag{1.46}$$

In the particular case $l = 0$, $q_l = -1$, we obtain the Schwarzschild solution in the prolate ellipsoidal coordinates:

$$\psi(x,y) = \frac{1}{2}\ln\frac{x-1}{x+1}, \quad \gamma(x,y) = \frac{1}{2}\ln\frac{x^2-1}{x^2-y^2}. \tag{1.47}$$

In conclusion, we can note that other coordinate systems are sometimes also used in obtaining static solutions of the Einstein equations.

Zipoy [36] considered oblate (flattened) ellipsoidal coordinates to obtain new solution of gravistatics. As was shown in Ref. [38], solutions thus obtained represent the gravitational field of a disk providing the Euclidean topology.

The solutions for a circular disk obtained in Ref. [39] have found application in cosmological hypotheses.

It should also be noted that vacuum solutions of the Einstein equations in toroidal coordinates were obtained by Misra [40] in 1961.

1.4 Method of Singular Sources

The right-hand side of (1.24) contains zero, though actually there should be a certain singular function characterizing the distribution of sources.

Let $\sigma(\rho, z)$ denote the mass density of such sources, and let us rewrite (1.24) in the form

$$\frac{1}{\rho}\frac{\partial}{\partial\rho}\left(\rho\frac{\partial\psi}{\partial\rho}\right) + \frac{\partial^2\psi}{\partial z^2} = -\sigma(\rho,z). \tag{1.48}$$

This equation has the solution

$$\psi = \frac{1}{4\pi} \int_v \frac{\sigma(\rho', z')}{|\vec{r} - \vec{r}'|} dv',$$

where $\vec{r}$ is the radius vector of the observation point and the primed coordinates describe the mass distribution. In the coordinates ρ, φ, z, we have

$$dv' = \rho' d\rho' d\varphi' dz',$$

$$|\vec{r} - \vec{r}'|^2 = \rho^2 + \rho'^2 - 2\rho\rho' \cos(\varphi - \varphi') + (z - z')^2.$$

Hence,

$$\psi(\rho, z) = \frac{1}{4\pi} \int_{\rho'=0}^{\infty} \int_{z'=-\infty}^{\infty}$$

$$\times \int_{\varphi'=0}^{2\pi} \frac{\sigma(\rho', z')\rho' d\rho' d\varphi' dz'}{\sqrt{\rho^2 + \rho'^2 - 2\rho\rho' \cos(\varphi - \varphi') + (z - z')^2}}. \tag{1.49}$$

Since the left-hand side of (1.49) does not depend on φ, we can set $\varphi = 0$ in the integral. If we choose

$$\sigma(\rho', z') = \frac{\delta(\rho' - \rho_0)}{\rho'} \sigma(z'),$$

where $\rho_0 = \text{const}$ and $\delta(\rho' - \rho_0)$ is Dirac's δ-function, we obtain

$$\psi(\rho, z) = \frac{1}{4\pi} \int_{z'=-\infty}^{\infty} \int_{\varphi'=0}^{2\pi} \frac{\sigma(z')d\varphi' dz'}{\sqrt{\rho^2 + \rho_0^2 - 2\rho\rho_0 \cos \varphi' + (z - z')^2}}. \tag{1.50}$$

First, let us examine some opportunities arising from the case $\rho_0 = 0$.

Example 1. Let $\sigma(z') = -\sigma^0(z')\theta(z')$, where $\theta(z')$ is the step function

$$\theta(z') = \begin{cases} 0 & \text{if } m < z' < -m \\ 1 & \text{if } -m < z' < m. \end{cases}$$

The solution (1.50) with these assumptions transforms to

$$\psi(\rho, z) = -\frac{1}{2} \int_{-m}^{m} \frac{\sigma^0(z')dz'}{\sqrt{\rho^2 + (z - z')^2}}. \tag{1.51}$$

Let $\sigma^0(z') = \delta_0 = $ const. With this choice, we come to the Zipoy solution:

$$\psi(\rho, z) = \frac{\delta_0}{2} \ln \left[\frac{z - m + \sqrt{\rho^2 + (z - m)^2}}{z + m + \sqrt{\rho^2 + (z + m)^2}} \right]. \qquad (1.52)$$

If we put $\delta_0 = 1$, then we obtain the Schwarzschild solution (1.35).

Example 2. Let $\sigma(z') = -2m\delta(z')$, where δ is Dirac's δ-function. Integration of (1.50) leads to

$$\psi_{\mathrm{Ch}} = -\frac{m}{\sqrt{\rho^2 + z^2}},$$

i.e., the Chazy–Curzon solution.

Example 3. Using the Chazy–Curzon solution, it is easy to obtain the static Euclidon solution:

$$\psi = -\frac{1}{2m} \int_0^z \psi_{\mathrm{Ch}}(\rho, z') dz' + \frac{1}{2} \ln \rho = \frac{1}{2} \ln(z + \sqrt{\rho^2 + z^2}). \quad (1.53)$$

This formal connection can be used for various generalizations. It follows from (1.53) that $\sigma(\rho, z) = 0$ for Euclidon solution.

We note that in the case when $\rho_0 = 0$ and $\sigma(z') = -2\sigma_1(z')$ are arbitrary functions, the solution of the gravistatic equations can be written in the form [41]

$$\psi(\rho, z) = -\int_{-\infty}^{\infty} \frac{\sigma_1(z')dz'}{R_1}, \qquad (1.54)$$

$$\gamma(\rho, z) = -\rho^2 \int_{-\infty}^{\infty} \int_{-\infty}^{\infty} \frac{\sigma_1(z')\sigma_1(z'')dz'dz''}{R_1 R_2[\rho^2 + R_1 R_2 + (z - z')(z - z'')]}, \qquad (1.55)$$

where

$$R_1 = \sqrt{\rho^2 + (z - z')^2}, \quad R_2 = \sqrt{\rho^2 + (z - z'')^2}.$$

Example 4. In the previous case, we assumed $\rho_0 = 0$. Let us $\rho_0 = $ const, $\sigma(z') = -\sigma^0(z')\theta(z')$, $\sigma^0(z') = 1$. Then, after integration in

(1.55) over z', we come to the following solution [42]:

$$\psi = \frac{1}{4\pi} \int_0^{2\pi} (\ln F_- - \ln F_+) d\varphi', \qquad (1.56)$$

where

$$F_\pm = z \pm m + \sqrt{(z \pm m)^2 - 2\rho\rho_0 \cos \varphi' + \rho^2 + \rho_0^2}.$$

If we put $\rho_0 = 0$, then (1.56) reduces to the Schwarzschild solution. In the case $\rho_0 = \text{const} \neq 0$, the solution (1.56) was investigated numerically. Calculations show that there is no connected horizon for any value of ρ_0.

Example 5. With the full elliptic integral of first kind $K(k)$ from (1.50), we obtain

$$\psi(\rho, z) = \frac{1}{\pi} \int_\infty^\infty dz' \frac{\sigma(z')K(k')}{\sqrt{(\rho + \rho_0)^2 + (z - z')^2}}, \qquad (1.57)$$

$$k' = \sqrt{\frac{4\rho\rho_0}{(\rho + \rho_0)^2 + (z - z')^2}}.$$

(a) Let $\sigma(z') = -2m\delta(z')$, where δ is Dirac's δ-function. Integration of (1.57) then leads to [43]

$$\psi = -\frac{2}{\pi} \frac{mK(k)}{\sqrt{(\rho + \rho_0)^2 + z^2}}, \quad k = \sqrt{\frac{4\rho\rho_0}{(\rho + \rho_0)^2 + z^2}}. \qquad (1.58)$$

In the case $\rho_0 = 0$, (1.58) leads to the Chazy–Curzon solution.

(b) Let $\sigma(z') = -\sigma^0(z')\theta(z')$, $\sigma^0(z') = 1$. With this choice from (1.57), we come to following solution:

$$\psi(\rho, z) = -\frac{1}{\pi} \int_{-m}^m \frac{K(k')dz'}{\sqrt{(\rho + \rho_0)^2 + (z - z')^2}}. \qquad (1.59)$$

Integration of (1.59) leads to

$$\psi(\rho, z) = \frac{1}{2} \ln \frac{z - m + \sqrt{(\rho + \rho_0)^2 + (z - m)^2}}{z + m + \sqrt{(\rho + \rho_0)^2 + (z - m)^2}}$$

$$- \frac{1}{2} \sum_{n=0}^{\infty} \left[\frac{(2n + 1)!!}{[2(n + 1)!!]} \right]^2 \left[\frac{4\rho\rho_0}{(\rho + \rho_0)^2} \right]^{n+1}$$

$$\times \sum_{r=0}^{n} \frac{(-1)^{r+1}}{2r + 1} \frac{n!}{(n - r)!r!}$$

$$\times \left[\left(\frac{z - m}{\sqrt{(\rho + \rho_0)^2 + (z - m)^2}} \right)^{2r+1} \right.$$

$$\left. - \left(\frac{z + m}{\sqrt{(\rho + \rho_0)^2 + (z + m)^2}} \right)^{2r+1} \right]. \qquad (1.60)$$

In the case $\rho_0 = 0$, we obtain from (1.60) the Schwarzschild solution.

In conclusion, we should mention the work of Bicak *et al.* [44–48], where the authors considered relativistic disks as sources of static vacuum spacetimes.

1.5 Waylen's Method

Waylen [49] has obtained a solution of (1.11) in the following integral form:

$$\psi(\rho, z) = \frac{1}{\pi} \int_0^{\pi} \left[f_1(u) + f_2(u) \ln \left(\frac{\rho}{\rho_0} \sin^2 \vartheta \right) \right] d\vartheta, \qquad (1.61)$$

where $f_1(u)$ and $f_2(u)$ are arbitrary functions of the argument $u = z + i\rho \cos \vartheta$. The coordinates ρ and z enter into the integral as parameters.

Since we are interested in the function ψ with finite values on the axis $\rho = 0$, we should put in (1.61) $f_2(u) = 0$ so that ψ is finally given by

$$\psi(\rho, z) = \frac{1}{\pi} \int_0^{\pi} f_1(u) d\vartheta. \qquad (1.62)$$

The use of the above expression is to some extend difficult because one should "guess" the form of the function $f_1(u)$ leading to physically relevant solutions.

For example, in the above-mentioned work, the only asymptotically flat solution, the Chazy–Curzon solution, was found. It follows from (1.62) with the choice of the function $f_1(u)$ in the form

$$f_1(u) = -\frac{m}{z + i\rho\cos\vartheta}.$$

As a result of integration, we have (1.26).

Assuming

$$f_1(u) = \frac{1}{2}\ln(z + w_0 + i\rho\cos\vartheta),$$

where w_0 is a constant of translation, we get the Euclidon solution.

According to the known relation

$$\int_0^\pi \ln(z + w_0 + i\rho\cos\vartheta)d\vartheta$$
$$= \pi\ln(z + w_0 + \sqrt{\rho^2 + (z + w_0)^2}) - \pi\ln 2,$$

we obtain the solution (1.32).

The Schwarzschild solution can be obtained by choosing the function $f_1(u)$ in the form

$$f_1(u) = \frac{1}{2}\ln\left(\frac{z - m + i\rho\cos\vartheta}{z + m + i\rho\cos\vartheta}\right),$$

and the integration reduces to the previous case.

1.6 Curzon-like Solutions

The linearity of the gravistatic equations makes it possible to solve the problem of superposition of two or several known solutions.

Bach and Weyl [43] were the first to formulate the problem of a two-particle solution in 1922. Further, the solution proposed by

Curzon,

$$\psi = -\frac{m_1}{R_1} - \frac{m_2}{R_2},$$

$$\gamma = -\frac{m_1^2 \rho^2}{2R_1^4} - \frac{m_2^2 \rho^2}{2R_2^4} + \frac{2m_1 m_2}{(z_1 - z_2)^2} \left[\frac{\rho^2 + (z - z_1)(z - z_2)}{R_1 R_2} - 1 \right],$$

$$(1.63)$$

where $R_{1,2} = \sqrt{\rho^2 + (z - z_{1,2})^2}$ and m_1, m_2, z_1, z_2 are constants, was interpreted (e.g., Ref. [50]) as a solution describing two particles of masses m_1 and m_2 at rest at the points z_1 and z_2.

In the particular case $m_1 = m_2 = m, z_1 = -z_2 = a$, the solution (1.63) is simplified:

$$\psi = -m \left(\frac{R_+ + R_-}{R_+ R_-} \right),$$

$$\gamma = \frac{m^2}{2a^2} \left[\frac{\rho^2 + (z^2 - a^2)}{R_+ R_-} - 1 \right] - \frac{m^2 \rho^2}{2R_+^4 R_-^4}(R_+^4 + R_-^4), \quad (1.64)$$

where $R_\pm = \sqrt{\rho^2 + (z \pm a)^2}$, and as $a \to 0$, the solution tends to the Chazy–Curzon solution.

Another combination of one-particle Curzon solutions was used in Ref. [51]:

$$\psi = -m \left(\frac{R_+ - R_-}{R_+ R_-} \right),$$

$$\gamma = -\frac{m^2}{2a^2} \left[\frac{\rho^2 + (z^2 - a^2)}{R_+ R_-} - 1 \right] - \frac{m^2 \rho^2}{2R_+^4 R_-^4}(R_+^4 + R_-^4). \quad (1.65)$$

This solution tends to the Schwarzschild metric as $a \to 0$. Consider Chazy–Curzon solution given by

$$\psi = -\frac{m}{\sqrt{\rho^2 + z^2}}. \qquad (1.66)$$

This potential can be used for obtaining some particular static solutions by passing from the canonical Weyl coordinates (ρ, z) to the prolate ellipsoidal coordinates (x, y).

Taking into account that equation (1.24) permits a shift of z to an arbitrary constants, a solution of the form

$$\psi = -\frac{m}{\sqrt{\rho^2 + (z \pm m)^2}} \tag{1.67}$$

follows from (1.66) immediately. Passing to the coordinates (x, y) according to (1.11), where one should put $k_0 = m$, we obtain from (1.67)

$$\psi = -\frac{1}{x \pm y}.$$

Bearing in mind the linearity of equation (1.16), one can construct from a static solution of the form

$$\psi = -\frac{ax + by}{x^2 - y^2}. \tag{1.68}$$

To do so, one should successively add and subtract the solutions $\psi = -1/(x + y)$ and $\psi = -1/(x - y)$ multiplying the results by the constants $(a - b)/2$ and $(a + b)/2$, respectively, and adding then the obtained expressions, one arrives at (1.68).

The solution (1.68) is of certain physical interest because it is asymptotically flat and exhibits physical properties similar to those of the Chazy–Curzon solution. Since this solution was not considered earlier, we give the corresponding expression for the metric function γ which, by integrating equations (1.17), (1.18), is found to be [52]

$$\gamma = -\frac{1 - y^2}{4(x^2 - y^2)^4}[2a^2 x^6 + 3(a^2 + 3b^2)x^4 y^2 - (a^2 + b^2)x^4$$

$$+ 2(2a^2 - b^2)x^2 y^4 - 6(a^2 + b^2)x^2 y^2 - (a^2 - b^2)y^6 - (a^2 + b^2)y^4$$

$$- 8abxy(x^2 - 1)(x^2 + y^2)]. \tag{1.69}$$

To examine the asymptotic behavior of the potential f, let us write it down in the curvature coordinates (r, θ) ($x = r/k_0 - 1$, $y = \cos\theta$, $k_0 = m$):

$$f = \exp\left[-\frac{2m(ar + bm\cos\theta - am)}{r^2 - 2mr + m^2\sin^2\theta}\right]. \tag{1.70}$$

Expanding f in inverse powers of r, we obtain the asymptotic of this solution:

$$f = 1 - 2\left[\frac{am}{r} - (a^2 - a - b\cos\theta)\frac{m^2}{r^2}\right.$$

$$+ \left(\frac{2}{3}a^3 - 2a^2 + a\cos^2\theta + a - 2ab\cos\theta + 2b\cos\theta\right)\frac{m^3}{r^3}$$

$$\left. + o\left(\frac{1}{r^4}\right)\right].$$

Consider now some particular cases of the solution.

Case 1. $a = 1, b = 0$. Then, for f and γ, we have

$$f = \exp\left(-\frac{2x}{x^2 - y^2}\right),$$

$$\gamma = -\frac{1 - y^2}{4(x^2 - y^2)^4}(2x^6 + 3x^4 y^2 - x^4 + 4x^2 y^4 - 6x^2 y^2 - y^4 - y^6),$$

$$(1.71)$$

and the asymptotic expansion of f is

$$f = 1 - 2\left[\frac{m}{r} + \frac{2}{3}P_2(\cos\theta)\frac{m^3}{r^3} + o\left(\frac{1}{r^4}\right)\right]. \tag{1.72}$$

Since (1.72) contains the monopole and quadrupole terms, one may consider the solution (1.71) as representing the gravitational field of a mass with a quadrupole moment equal to $(2/3)m^3$.

Case 2. $a = 1, b = \pm 1$. In this case, we obtain from

$$f = \exp[-2/(x \mp y)], \quad \gamma = -\frac{(x^2 - 1)(1 - y^2)}{2(x \pm y)^4}. \tag{1.73}$$

From it follows that

$$f = 1 - 2\left[\frac{m}{r} \pm P_1(\cos\theta)\frac{m^2}{r^2} + \frac{2}{3}P_2(\cos\theta)\frac{m^3}{r^3} + o\left(\frac{1}{r^4}\right)\right]. \tag{1.74}$$

In addition to the monopole and quadrupole terms, the expansion (1.74) also contains dipole terms. This particular case of the solution

(1.68) has been used in Ref. [53] for the construction of electrostatic solutions of the Einstein–Maxwell equations.

Case 3. $a = 0$. In this case, we have

$$f = \exp[-2by/(x^2 - y^2)],$$

$$\gamma = -\frac{b^2(1 - y^2)}{4(x^2 - y^2)^4}(9x^4y^2 - x^4 - 2x^2y^4 - 6x^2y^2 - y^4 + y^6), \quad (1.75)$$

and the asymptotic behavior of f is determined by the expression

$$f = 1 - 2\left[bP_1(\cos\theta)\frac{m^2}{r^2} + 2bP_2(\cos\theta)\frac{m^3}{r^3} + o\left(\frac{1}{r^4}\right)\right]. \quad (1.76)$$

In the expression (1.76), monopole terms are absent, and we only have the dipole term corresponding to the required Newtonian limit. This property (1.75) of the solution allows it to be used in generating other solutions with definite asymptotic behavior.

1.7 Superposition of Static Soliton Solutions

Belinsky and Zakharov [54, 55] found a 2n-soliton solution of the stationary Einstein equations in an analytic form by the inverse scattering method.

Using Belinski and Zakharov's method, Chaudhiri and Das [56] obtained a new odd-soliton solution in a determinant form and showed to that it includes Weyl's half-integral delta static solutions in a special case.

Alekseev and Belinsky [57] studied the structure of singularities and horizons of static 2n-soliton metrics:

$$ds^2 = \frac{\mu_1 \cdots \mu_{2n}}{\rho^{2n}}dt^2 - \frac{\rho^{2n}}{\mu_1 \cdots \mu_{2n}}\left\{C_0\rho^{6n^2}\left(\frac{\mu_1 \cdots \mu_{2n}}{\rho^{2n}}\right)^{2n}\right.$$

$$\left. \cdot [\Pi^{2n}_{k,l=1}(\rho^2 + \mu_k\mu_l)]^{-1}(d\rho^2 + dz^2) + \rho^2 d\varphi^2\right\}, \quad (1.77)$$

where

$$\mu_k = W_k - z + \varepsilon_k[(W_k - z)^2 + \rho^2]^{1/2}, \quad k = 1, 2, \ldots, 2n.$$

To ensure that metric (1.77) describes asymptotically flat fields, the condition must be satisfied

$$\varepsilon_k = \pm 1, \quad \sum_{k=1}^{2n} \varepsilon_k = 0.$$

W_k are related to the mass of the sources in a specific way. The function μ_k/ρ is called a gravitational soliton. It is easy to verify that (1.77) is a solution to equation (1.11).

Two-soliton solution

$$\psi = \frac{1}{2} \ln \frac{W_1 - z + \sqrt{\rho^2 + (W_1 - z)^2}}{W_2 - z + \sqrt{\rho^2 + (W_2 - z)^2}}. \tag{1.78}$$

As $[W_1] \to [W_2]$, the solution transitions into the Schwarzschild solution (a single-center solution). Increasing the number of soliton pairs allows for the construction of a multicenter solution. These gravitational centers appear to be strung along a single axis, forming a pattern for a distant observer that resembles a string of beads.

The quantities μ_k can be written in a different form if, instead of the real values W_k, complex ones are used:

$$W_k = z_k + i\sigma_k. \tag{1.79}$$

Then,

$$\mu_k = r_k e^{i\gamma_k}, \tag{1.80}$$

where

$$r_k = R_k + \varepsilon_k \sqrt{R_k^2 - \rho^2},$$

$$R_k = \frac{1}{2}\sqrt{(z_k - z)^2 + (\rho - \sigma_k)^2} + \frac{1}{2}\sqrt{(z_k - z)^2 + (\rho + \sigma_k)^2},$$

$$\cos\gamma_k = \frac{\varepsilon_k(z_k - z)}{\sqrt{R_k^2 - \rho^2}}, \quad \sin\gamma_k = \frac{\sigma_k}{R_k}. \tag{1.81}$$

1.8 Physical Interpretation of One-Static Euclidon Solution

The progress in the development of general relativity (GR) and in understanding its physical content has largely been determined by exact solutions to Einstein's equations. For this reason, the problem of obtaining and studying exact solutions in GR is becoming increasingly important. It cannot be said that the analysis of any particular exact solution has been extensively covered in publications. As early as 1975, Kinnersley wrote, "Most known solutions describe blatantly unphysical situations, and there is a tendency to pay the least attention to the most useful solutions."

In this section, the physical content of static Euclidon solutions to Einstein's equations is analyzed. These solutions serve as "building blocks" of the theory, which allows for the construction of almost all known solutions to the vacuum static axially symmetric Einstein equations, including such important ones as the Schwarzschild solution.

One of the solutions to equations (1.11), (1.12) has the form

$$f = \frac{z - z_1 + \sqrt{\rho^2 + (z - z_1)^2}}{C_1'},$$

$$\gamma = \frac{1}{2} \ln \frac{z - z_1 + \sqrt{\rho^2 + (z - z_1)^2}}{\sqrt{\rho^2 + (z - z_1)^2}} + \gamma_0', \tag{1.82}$$

where z_1, C_1', γ_0' are constant values. The solution is the so-called static Euclidon solution.

According to this solution, the corresponding metric is

$$ds^2 = C_1' e^{2\gamma_0'} \frac{d\rho^2 + dz^2}{\sqrt{\rho^2 + (z - z_1)^2}} + \frac{C_1' \rho^2 d\varphi^2}{z - z_1 + \sqrt{\rho^2 + (z - z_1)^2}}$$

$$- \frac{z - z_1 + \sqrt{\rho^2 + (z - z_1)^2}}{C_1'} c^2 dt^2. \tag{1.83}$$

It is easy to see that the transformation [83]

$$2C_1' e^{2\gamma_0'} \rho = \rho' \sqrt{(z' - z_1')^2 - c^2 t'^2},$$

$$\varphi = \varphi' e^{\gamma_0'} \sqrt{2},$$

$$4C_1' e^{2\gamma_0'} (z - z_1) = (z' - z_1')^2 - c^2 t'^2 - \rho'^2,$$

$$\frac{e^{-\gamma_0'} ct}{C_1' \sqrt{2}} = \frac{1}{2} \ln \frac{z' - z_1' + ct'}{z' - z_1' - ct'} \tag{1.84}$$

transforms the metric into the Minkowski metric

$$ds_M^2 = d\rho'^2 + \rho'^2 d\varphi'^2 + dz'^2 - c^2 dt'^2. \qquad (1.85)$$

Thus, the metric of the static Euclidon solution describes flat euclidean spacetime.

The formulas for the inverse transformation are as follows:

$$\rho' = e^{\gamma_0'}\sqrt{2C_1'\mu_-},$$

$$\varphi' = \frac{e^{-\gamma_0'}}{\sqrt{2}}\varphi,$$

$$z' - z_1' = \varepsilon e^{\gamma_0'}\sqrt{2C_1'\mu_+}\cosh\frac{e^{-\gamma_0'}ct}{C_1'\sqrt{2}},$$

$$ct' = e^{\gamma_0'}\sqrt{2C_1'\mu_+}\sinh\frac{e^{-\gamma_0'}ct}{C_1'\sqrt{2}}, \qquad (1.86)$$

where

$$\mu_\pm = \sqrt{\rho^2 + (z - z_1)^2} \pm (z - z_1), \quad \varepsilon = \pm 1.$$

Returning to the solution, it can be shown through straightforward but cumbersome calculations that this solution reduces all components of the Riemann curvature tensor R_{iklm} to zero. Therefore, the metric remains within the framework of special relativity (SR).

This fact is the reason for the name "Euclidon."

All of the above indicates that the Euclidon solution is associated with a relativistic reference frame and has a clear physical interpretation.

From now on, we refer to the primed reference frame $(\rho', \varphi', z', t')$ as the stationary or inertial reference frame (IRF) and (ρ, φ, z, t) as the non-inertial reference frame (NIRF). The transformation between the two is carried out using specific formulas.

Now, let's consider a point at rest in the non-inertial system (ρ, φ, z, t):

$$\rho = \rho_0, \quad z = z_0, \quad \varphi = \varphi_0 e^{\gamma_0'}\sqrt{2}. \qquad (1.87)$$

For it,

$$\mu_{\pm}^0 = \sqrt{\rho_0^2 + (z_0 - z_1)^2} \pm (z_0 - z_1).$$
(1.88)

We also introduce the notations

$$z' = \frac{c^2}{a_0'}, \quad a_0' = \varepsilon \frac{c^2}{e^{\gamma_0}\sqrt{2C_1'\mu_+^0}}.$$
(1.89)

From the perspective of the inertial reference frame (IRF), this point is in motion. Its coordinates, relativistic velocity, and acceleration can be easily computed:

$$\rho' = e^{\gamma_0}\sqrt{2C_1'\mu_-^0} = \text{const} = \rho_0',$$

$$\varphi' = \varphi_0,$$

$$z' = \frac{c^2}{a_0'}\left[\sqrt{1 + \left(\frac{a_0't'}{c}\right)^2} - 1\right],$$

$$v' = \frac{dz'}{dt'} = \frac{a_0't'}{\sqrt{1 + \left(\frac{a_0't'}{c}\right)^2}},$$

$$a' = \frac{d}{dt'}\left[\frac{v'}{\sqrt{1 - \frac{v'^2}{c^2}}}\right] = a_0'.$$
(1.90)

The motion of the point, according to formulas (1.90), is rectilinear relativistic uniformly accelerated motion along the z' axis. It is sometimes referred to as hyperbolic.

For (1.90), when $\rho_0 = 0$, we will consider three particular cases:
(1)

$$\rho_0 = 0, \quad z_1 = m, \quad z_0 > m,$$

$$z' = 2\varepsilon e^{\gamma_0}\sqrt{C_1'(z_0 - m)}\left[\sqrt{1 + \frac{c^2 t'^2}{4e^{2\gamma_0}C_1'(z_0 - m)}} - 1\right],$$

$$v_-' = \frac{a_0'^- t'}{\sqrt{1 + \left(\frac{a_0't'}{c}\right)^2}}, \quad a_0'^- = \varepsilon\frac{c^2}{2e^{\gamma_0}\sqrt{C_1'(z_0 - m)}}.$$
(1.91)

(2)

$$\rho_0 = 0, \quad z_1 = -m, \quad z_0 > -m,$$

$$z' = 2\varepsilon e^{\gamma_0'} \sqrt{C_1'(z_0 + m)} \left[\sqrt{1 + \frac{c^2 t'^2}{4e^{2\gamma_0'} C_1'(z_0 + m)}} - 1 \right],$$

$$v_+' = \frac{a_0'^+ t'}{\sqrt{1 + \left(\frac{a_0' t'}{c}\right)^2}}, \quad a_0'^+ = \varepsilon \frac{c^2}{2e^{\gamma_0'} \sqrt{C_1'(z_0 + m)}}. \tag{1.92}$$

(3) For $\rho_0 = 0$ and $z_0 \le z_1$, the acceleration a_0 is not defined, and instead of (1.90), we obtain from (1.89)

$$\rho' = 2e^{\gamma_0'} \sqrt{C_1'(z_1 - z_0)}, \quad \varphi' = \varphi_0, \quad z' = z_1'. \tag{1.93}$$

In this region of space, all points in the inertial reference frame (IRF) are at rest.

1.9 Superposition of Static Euclidon Solutions

As was shown above, the one-Euclidon solution

$$\psi(\rho, z) = \frac{1}{2} \ln(z + w + \sqrt{\rho^2 + (z + w)^2}),$$

$$\gamma = \frac{1}{2} \ln \left(\frac{z + w + \sqrt{\rho^2 + (z + w)^2}}{\sqrt{\rho^2 + (z + w)^2}} \right) \tag{1.94}$$

satisfies equations (1.11), (1.12) and corresponds to Euclidean space. Here, w is a constant of translation. Superposition of several solutions of type (1.94) with different translation constants is of interest for searching exact solution.

Due to linearity of equation (1.11), a linear superposition of solutions (1.94) with different translation constants w is also a solution of the above equations:

$$\psi(\rho, z) = \frac{1}{2} \sum_{i=1}^{2N} \varepsilon_i \ln(z \pm w_i + \sqrt{\rho^2 + (z \pm w_i)^2}). \tag{1.95}$$

We further use the terminology of Ref. [57]

Only even-number Euclidon solutions (1.95) correspond to asymptotically flat solutions. The constants ε_i should satisfy the following conditions:

1. $\varepsilon_i = \pm 1$,

2. $\sum_{i=1}^{2N} \varepsilon_i = 0$.

In the general case, the number of solutions satisfying the asymptotic flatness conditions can be given by

$$S = (2N)!/(N!)^2,$$

where N is the number of Euclidon pairs.

For convenience, to analyze $2N$-Euclidon solution, one can consider geometric mapping of Euclidons with the help of the constants w_i connecting them with the moving Weyl coordinate z.

The behavior of the metric function f near a point on the z axis permits one to understand if this point is regular or it belongs to a horizon ($f \to 0$) or coordinate singularity ($f \to \infty$). This, in turn, allows one to describe the structure of the source of the gravitational field and thereby to give a physical interpretation to the solution:

1. For the case of two-euclidean solutions ($N = 1$), there are two types of diagrams:

1-I. The solution ($w_1 \leq w_2$)

$$\psi(\rho, z) = \frac{1}{2} \ln \left(\frac{z + w_1 + \sqrt{\rho^2 + (z + w_1)^2}}{z + w_2 + \sqrt{\rho^2 + (z + w_2)^2}} \right) \qquad (1.96)$$

has a connect horizon, and for $w_1 = -w_2 = -m$ (where m is the mass of a source of the gravitational field), it turns into the Schwarzschild solution (1.35).

1-II. This case corresponds to a solution with a coordinate singularity which can be obtain from by changing the sign of the mass (1.96). Dynamical system consisting of two equal and opposite masses were studied by Bondi [58], Alekseev and Belinsky [57] and a superposition of two Schwarzschild solutions can be found in Ref. [59].

2. In the case of a four-Euclidean solution ($N = 2$), we have six types of diagrams ($w_1 \leq w_2 \leq w_3 \leq w_4$).

2-I. The solution

$$\psi(\rho, z) = \frac{1}{2} \ln \left(\frac{z + w_1 + \sqrt{\rho^2 + (z + w_1)^2}}{z + w_3 + \sqrt{\rho^2 + (z + w_3)^2}} \right)$$

$$+ \frac{1}{2} \ln \left(\frac{z + w_2 + \sqrt{\rho^2 + (z + w_2)^2}}{z + w_4 + \sqrt{\rho^2 + (z + w_4)^2}} \right) \quad (1.97)$$

has a connect horizon, and when $w_1 = w_2 = -m, w_3 = w_4 = m$, we arrive at the Darmois solution.

2-II. The solution

$$\psi(\rho, z) = \frac{1}{2} \ln \left(\frac{z + w_1 + \sqrt{\rho^2 + (z + w_1)^2}}{z + w_2 + \sqrt{\rho^2 + (z + w_2)^2}} \right)$$

$$+ \frac{1}{2} \ln \left(\frac{z + w_3 + \sqrt{\rho^2 + (z + w_3)^2}}{z + w_4 + \sqrt{\rho^2 + (z + w_4)^2}} \right) \quad (1.98)$$

has two horizons with a region of regular behavior of the gravitational potential between them.

It is interesting to note that if $w_2 \to w_3$, then in the limit, we arrive at the Schwarzschild solution (1.35).

It means that the four-Euclidean solution 2-II formally represents the gravitational field of two black holes at rest at constant distance. Nevertheless, this is in contradiction with the law of physical bodies' motion according to which these black holes should fall onto each other. This fact needs no Euclidean locality of the space metric as is in the case of known Chazy–Curzon solution.

2-III. The solution has a horizon and a coordinate singularity corresponding to a negative mass of the source.

Other cases of the four-euclidean solution are of no interest since they can be reduced to previous case. For example, the case 2-IV can be obtained from 2-III by the transformation $z \to -z$, while 2-V and 2-VI are analogues of 2-I and 2-II with negative masses.

Multiparticle systems were studied in Refs. [15, 50]. Basing on these results, Israel and Khan [61] (see also Ref. [62]) obtained in

1964 a metric corresponding to N collinear non-crossing bars (rods, pivots) with masses multiple of $\pm 1/2$. Hence, a $2N$-euclidean solution can be written in the form

$$\psi = \frac{1}{2} \sum_{i=1}^{N} \ln \left(\frac{R_{i+} + R_{i-} - 2w_i}{R_{i+} + R_{i-} + 2w_i} \right),$$

$$\gamma = \frac{1}{4} \sum_{i=1}^{N} \sum_{j=1}^{N} \ln \left(\frac{E(i_+, j_-)E(i_-, j_+)}{E(i_-, j_-)E(i_+, j_+)} \right), \tag{1.99}$$

where

$$R_{i\pm} = \sqrt{\rho^2 + (z \pm w_i)^2}$$

and

$$E(i_+, j_-) = E(j_-, i_+) = \rho^2 + R_{i+}R_{j+} + (z + w_i)(z - w_j).$$

The other $E(i_-, j_+), E(i_-, j_-)$ and $E(i_+, j_+)$ are defined in a similar way.

In case $w_i = m$ in (1.99), we can obtain the Zipoy solution with the parameter $\delta_0 = N$.

In conclusion, we note that a superposition of N Schwarzschild black holes was discussed in Refs. [63–65], and the conditions of stability can be found in Ref. [25]. Reference [66] is devoted to a superposition of the Zipoy solution with Weyl class metrics.

1.10 Some Physical Interpretation of Two-Static Euclidon Solution (Schwarzschild Solution)

1. The Schwarzschild metric in curvature coordinates has the form

$$ds_{\text{Sch.}}^2 = \frac{dr^2}{g_{00}} + r^2(d\theta^2 + \sin^2\theta d\varphi^2) - g_{00}c^2 dt^2, \tag{1.100}$$

where $g_{00} = 1 - \frac{2m}{r}$.

Let's consider a small neighborhood around a fixed value of $r = r_0$.

Within this neighborhood, we can represent the metric in the form

$$g_{00}(r) \approx a + b(r - r_0), \quad a = g_{00}(r_0) = 1 - \frac{2m}{r_0}, \quad b = \frac{2m}{r_0^2},$$

$$(1.101)$$

and by using transformations, we can obtain [67]

$$r = r_0 - \frac{a}{b} + \frac{b}{4}[(R - R_1')^2 - c^2 T^2],$$

$$ct = \frac{2}{b} \operatorname{arctanh} \frac{cT}{R - R_1'},$$

$$R - R_1' > cT, \qquad (1.102)$$

or

$$r = r_0 - \frac{a}{b} - \frac{b}{4}[c^2 T^2 - (R - R_1')^2],$$

$$ct = \frac{2}{b} \operatorname{arctanh} \frac{R - R_1'}{cT},$$

$$R - R_1' < cT. \qquad (1.103)$$

We transition to any region of the 4-space to a reference frame, which we call the quasi-inertial reference frame (QIRF), with the metric

$$ds_{\text{Sch.}}^2 \approx dR^2 + r^2(d\theta^2 + \sin^2 \theta d\varphi^2) - c^2 dt^2. \qquad (1.104)$$

As $r \sim R$, the metric approaches the Galilean metric.

The inverse transformation formulas are as follows:

$$R - R_1' = \varepsilon \frac{2}{\sqrt{b}} \sqrt{r - r_0 + \frac{a}{b}} \cosh \frac{bct}{2} \approx \varepsilon \frac{2}{b} \sqrt{g_{00}} \cosh \frac{bct}{2},$$

$$cT = \frac{2}{\sqrt{b}} \sqrt{r - r_0 + \frac{a}{b}} \sinh \frac{bct}{2} \approx \frac{2}{b} \sqrt{g_{00}} \sinh \frac{bct}{2},$$

$$\varepsilon = \pm 1,$$

$$R - R_1' > cT. \qquad (1.105)$$

At $t = 0$, it follows that

$$\varepsilon dR \approx \frac{dr}{\sqrt{g_{00}}},$$

$$dT \approx \sqrt{g_{00}} dt.$$

Let's consider a stationary point at $r = r_0$ in spacetime. From the perspective of the quasi-inertial reference frame (QIRF), it undergoes motion

$$R = R_1 + \frac{c^2}{\alpha_0}\left[\sqrt{1 + \left(\frac{\alpha_0 T}{c}\right)^2} - 1\right],$$

$$V = \frac{dR}{dT} = \frac{\alpha_0 T}{\sqrt{1 + \left(\frac{\alpha_0 T}{c}\right)^2}},$$

$$a = \frac{d}{dt}\left[\frac{V}{\sqrt{1 - V^2/c^2}}\right] = \alpha_0. \tag{1.106}$$

Here, we have chosen $R'_1 = -\frac{c^2}{\alpha_0} + R_1$, $\varepsilon = -1$, $R_1 \approx r_0$ and the acceleration is given by

$$\frac{\alpha_0}{c^2} \approx -\frac{m}{r_0^2 \sqrt{g_{00}}}. \tag{1.107}$$

2. The static Euclidon solution has a clear physical interpretation as a relativistic accelerated non-inertial reference frame, which provides a different perspective on the physical interpretation of well-known solutions, such as the Schwarzschild solution.

It is easy to see that the Schwarzschild metric (1.35), (1.36) is approximately equal to the metric of the static Euclidon solution (1.83) in a small region of spacetime $\rho \approx \rho_0$, $z \approx z_0$ if we assume that

$$\begin{cases} C'_1 = z_0 + m + \sqrt{\rho_0^2 + (z_0 + m)^2} \\[2ex] e^{2\gamma'_0} = \dfrac{(\sqrt{\rho_0^2 + (z_0 + m)^2} + \sqrt{\rho_0^2 + (z_0 - m)^2})^2 - 4m^2}{4\sqrt{\rho_0^2 + (z_0 + m)^2}(z_0 - m + \sqrt{\rho_0^2 + (z_0 - m)^2})}. \end{cases} \tag{1.108}$$

By a coordinate transformation (1.86) with $z_1 = m$, the Schwarzschild metric in this region of spacetime can be approximately transformed into the Minkowski metric.

With this approach, the Schwarzschild metric in a small region of spacetime can be approximately regarded as a quasi-non-inertial reference frame (QNIRF).

2-I. Let us consider a small neighborhood of the region within this non-inertial reference frame:

$$\rho = \rho_0 = 0, \quad z = z_0 = r_0 - m, \quad \varphi = \varphi_0$$

or

$$r = r_0, \quad \theta = 0, \quad \varphi = \varphi_0, \tag{1.109}$$

where (r, θ) are curvature coordinates.

In this case, we obtain from (1.86)

$$\begin{cases} \rho' = 0 \\[2mm] \varphi' = \varphi_0 \\[2mm] z' - z'_1 = -2\sqrt{r_0}\sqrt{r - 2m}\cosh\dfrac{t}{2r_0} \\[3mm] t' = 2\sqrt{r_0}\sqrt{r - 2m}\sinh\frac{t}{2r_0}. \end{cases} \tag{1.110}$$

At $t = 0$, it is easy to see that
$dt' \approx dt\sqrt{1 - \frac{2m}{r}}$ is the true time and
$-dz' \approx \dfrac{dr}{\sqrt{1 - \frac{2m}{r}}}$ is the true distance between two nearby points
along r in Schwarzschild spacetime (e.g., Ref. [70]).

From the perspective of the quasi-inertial reference frame (QIRF), a stationary point $\rho = \rho_0 = 0$, $z = z_0$, $\varphi = \varphi_0$ undergoes relativistic (hyperbolic) uniformly accelerated motion along the z' axis.

$$\begin{cases} \rho' = 0 \\[2mm] \varphi' = \varphi_0 \\[2mm] z' = \dfrac{1}{a'_0}\left[\sqrt{1 + (a'_0 t')^2} - 1\right] + z'_0 \\[3mm] v' = \dfrac{dz'}{dt'} = \dfrac{a'_0 t'}{\sqrt{1 + (a'_0 t')^2}} \\[3mm] a' = \dfrac{d}{dt'}\left[\dfrac{v'}{\sqrt{1 - v'^2}}\right] = a'_0. \end{cases} \tag{1.111}$$

Here, we have assumed $z_1' = -\frac{1}{a_0'} + z_0'$, $\varepsilon = -1$ and the acceleration has the form

$$a_0' = -\frac{1}{2r_0\sqrt{1 - \frac{2m}{r_0}}}; \quad |a_0'| \to \infty, \quad r_0 \approx r \to 2m. \tag{1.112}$$

Then, approximately for $r \approx r_0 \to \infty$, $z_0' \approx z_0$, we can write

$$\frac{a_0'}{\sqrt{g_{00}}} \approx -\frac{1}{2r}\left(1 + \frac{2m}{r} + \cdots\right). \tag{1.113}$$

Similarly, as in the Newtonian limit where terms with $\frac{2m}{r}$ are retained, in this case, we can obtain

$$-\delta r \approx \frac{1}{\alpha}\left[\sqrt{1 + (\alpha\delta t)^2} - 1\right], \quad \alpha = -\frac{m}{r^2},$$

which, considering spherical symmetry, corresponds to radial geodesics (falling of a particle toward the center).

Indeed, it follows from (1.110) that

$$d\tau^2 = dt'^2 - dz'^2 = \frac{r}{r_0}g_{00}(r)dt^2 - \frac{r_0}{r}\frac{dr^2}{g_{00}(r)}. \tag{1.114}$$

Using the connection of own time τ and Schwarzschild's time t (e.g., Ref. [71]),

$$dt = \frac{\sqrt{g_{00}(r_i)}}{g_{00}(r)}d\tau, \tag{1.115}$$

such that $r = r_i, \dot{r} = 0$.

From (1.114), we get

$$\left(\frac{dr}{d\tau}\right)^2 = \frac{2m}{r}\cdot\left(\frac{r}{r_0}\right) - \frac{2m}{r_i}\cdot\left(\frac{r}{r_0}\right)^2 - \frac{r}{r_0}\left(1 - \frac{r}{r_0}\right). \tag{1.116}$$

Since $m = M$, $r \approx r_0$, we obtain [71]

$$\left(\frac{dr}{d\tau}\right)^2 \approx \frac{2M}{r}\cdot\left(1 - \frac{r}{r_i}\right). \tag{1.117}$$

At $r_i \approx r_0 \approx r$, it is easy to see that

$$dt' \approx d\tau\cosh\frac{t}{2r_0}$$

or

$$d\tau \approx dt'\sqrt{1 - v'^2}.$$

The case for (1.93) corresponds to the region inside the event horizon $r < 2m$.

2-II. One can consider the case where the point moves along the radius r toward the center in Schwarzschild spacetime. In this case, one can arrive at a spiral motion toward the gravitating center:

$$r' = r'(\cos\theta'), \quad \varphi' = \text{const.}$$

In the quasi-inertial reference frame (QIRF), gravitational capture is possible.

Also, the spiral motion in QIRF toward the gravitating center

$$\rho' = \rho'(\varphi'), \quad z' = 0$$

can be formal obtained from (1.86), (1.108) by setting

$$-z_1'(r_0) = e^{\gamma_0'}\sqrt{2C_1'\mu_+}\cosh\frac{e^{-\gamma_0't}}{C_1'\sqrt{2}} = \frac{C_1'\sqrt{2}}{e^{-\gamma_0'}}\sqrt{g_{00}}. \tag{1.118}$$

Thus, we obtain

$$\rho_0 \approx \rho(t) = \sqrt{r(r - 2m)}, \quad z = z_0 = 0, \quad \varphi = \varphi(r_0),$$

$$r_0 \approx r = r(t) = r_0 e^{-4\gamma_0'}\cosh^{-2}\frac{e^{-\gamma_0't}}{r_0\sqrt{2}}, \quad \theta = \frac{\pi}{2}, \tag{1.119}$$

where

$$e^{2\gamma_0'} = \frac{r_0}{r_0 - m}.$$

In this case, we have

$$\rho' = \sqrt{r_0 r}\frac{\varphi(r_0)}{\varphi'},$$

$$z' = 0,$$

$$t' = \frac{r_0\sqrt{2}}{e^{-\gamma_0'}}\sqrt{g_{00}}\tanh\frac{e^{-\gamma_0't}}{r_0\sqrt{2}}. \tag{1.120}$$

At $t = 0$, it is easy to see that

$$dt' \approx dt\sqrt{g_{00}}.$$

Then, approximately for $r \approx r_0 \to \infty$,

$$\delta r \approx \frac{1}{2}\alpha(\delta\tau)^2.$$

1.11 Geroch–Hansen Multipole Moments and the Erez–Rosen Solution

Hoenselaers [70] proposed a simplified procedure of calculating the relativistic Geroch–Hansen multipole moments [68, 69]. In accord with this method, we consider f potential of the form

$$\xi(x,y) = \frac{1-f}{1+f} \tag{1.121}$$

in prolate ellipsoid coordinates such that $y = 1$ on the symmetry axis (the case $y = -1$ provides similar results).

Introducing the canonical Weyl coordinate $z = k_0 xy$, one can define the infinite point Λ as a limit $z \to \infty$ or $\tilde{z} \to 0$ for the inverse coordinate

$$\tilde{z} = 1/z, \tag{1.122}$$

then Geroch–Hansen multipole moments for static solutions are defined by the transformed potential

$$\tilde{\xi}(\tilde{z},1) = \frac{1}{\tilde{z}}\xi(\tilde{z},1) \tag{1.123}$$

in the form

$$M_l = m_l + d_l(m_{l-1},\ldots,m_0), \tag{1.124}$$

where the first six dipole terms d_l have the form

$$d_0 = d_1 = d_2 = d_3 = 0, \quad d_4 = \frac{1}{7}m_0(m_1^2 - m_0 m_2),$$

$$d_5 = \frac{1}{3}m_0(m_1 m_2 - m_0 m_3) + \frac{1}{21}m_1(m_1^2 - m_0 m_2) \tag{1.125}$$

and

$$m_l = \frac{1}{l!}\left(\frac{d^l\tilde{\xi}(\tilde{z},1)}{d\tilde{z}^l}\right)_{\tilde{z}=0}. \tag{1.126}$$

Therefore, in order to find the relativistic multipole moments, one should calculate the values m_l.

Expanding in powers of $\tilde{z}$ the potential $\xi(\tilde{z},1)$

$$\xi(\tilde{z},1) = \sum_{k=1}^{\infty}\left(\frac{d^k\xi(\tilde{z},1)}{d\tilde{z}^k}\right)_{\tilde{z}=0}\frac{\tilde{z}^k}{k!}, \tag{1.127}$$

we obtain from (1.123) and (1.126)

$$m_l = \frac{1}{(l+1)!}\left(\frac{d^{l+1}\xi(\tilde{z},1)}{d\tilde{z}^{l+1}}\right)_{\tilde{z}=0}. \tag{1.128}$$

Substitution of (1.121) into (1.128) gives the following recursion relation for m_l:

$$m_l = \left(-\frac{h_{l+1}}{(l+1)!}\right)_{y=1,\tilde{z}=0}, \tag{1.129}$$

where

$$h_1 = \frac{1}{2f}\frac{df}{d\tilde{z}}, \tag{1.130}$$

$$h_l = \frac{dh_{l-1}}{d\tilde{z}} + 2\xi h_1 h_{l-1}, l \geq 2. \tag{1.131}$$

Hence, the procedure of finding Geroch–Hansen multipole moments is as follows:

(1) Calculate the potentials ξ according to (1.121).
(2) Find the quantities m_l according to (1.129), (1.130) and (1.131).
(3) Obtain the mass multipole moments M_l according to (1.124), where d_l is related to m_l by (1.125).

Note that relativistic and coordinate-invariant calculations of the multipole moments were studied in Refs. [71–73], but in spite of various mathematical methods, they are physically equivalent [72–74].

1. As was shown in Section 1.3, item 7, an asymptotically flat solution of equation (1.16) can be written in the form

$$\psi(x,y) = \sum_{l=0}^{\infty} q_l Q_l(x) P_l(y), \tag{1.132}$$

where $P_l(y)$ are the Legendre polynomials and $Q_l(x)$ are the Legendre function of the second type.

The particular solution of the form

$$\psi(x,y) = \frac{1}{2}\ln\frac{x-1}{x+1} + q_l Q_l(x)P_l(y) \tag{1.133}$$

was interpreted by Erez and Rosen as the gravitational field of a particle endowed by a mass 2_l-dipole moment.

Quevedo [76] points out the the necessity of a coordinate-invariant study of the Newtonian limit by the Ehlers method [74]. According to this method, Newtonian potential Φ of a given static axially symmetric vacuum solution can found by calculating the limit

$$\Phi = \lim_{\lambda\to 0}\frac{\psi(\rho,z,\lambda)}{\lambda}, \tag{1.134}$$

where $\lambda = 1/c^2$ (c is the velocity of light) and $\psi(\rho,z,\lambda)$ is the metric function in canonical Weyl coordinates including X in an explicit form.

This is done by transforming to the CGS system by changing the parameter m to $MG\lambda$, where M is the gravitational source mass in the new system of units and G is the gravitational constants.

For instance, using the Ehlers method for the Schwarzschild solution, one can easily find that

$$\Phi = -\frac{GM}{\sqrt{\rho^2 + z^2}}, \tag{1.135}$$

i.e., it coincides with the Newtonian potential of a spherically symmetric distribution of mass.

Quevedo [77] considered the generalized Erez–Rosen solution defined by the potential

$$\psi = \frac{1}{2}\ln\left(\frac{x-1}{x+1}\right) + \sum_{l=0}^{\infty}(-1)^{l+1}q_l Q_l(x)P_l(y). \tag{1.136}$$

He has shown than this solution represents a superposition of the Schwarzschild metric

$$\psi = \frac{1}{2}\ln\left(\frac{x-1}{x+1}\right)$$

with an arbitrary set of mass multipole moments determined by the constants q_l. In Ref. [77], he integrated equations (1.17), (1.18) with the potential (1.136). The corresponding metric function γ of the generalized Erez–Rosen solution can be obtained by some tedious calculations.

The virtue of the generalized Erez–Rosen solution is the fact that it contains the Schwarzschild solution in an explicit form. Therefore, it permits considering arbitrary deviations of the source from spherical symmetry. However, in Ref. [78], it was shown that in the solution the multipoles ψ_l make the Schwarzschild horizon defined by the hypersurface $x = 1$ singular. It means that no test body can cross the horizon.

2. The Newtonian potentials for solution (1.136), calculated according to, have the form

$$\Phi = G \sum_{n=0}^{\infty} (-1)^{n+1} \frac{n!}{(2n+1)!!} \tilde{q}_n M^{n+1} \frac{P_n(\cos\vartheta)}{R^{n+1}}, \qquad (1.137)$$

where $\cos\vartheta = \frac{z}{R}$, $R = \sqrt{\rho^2 + z^2}$ and $\tilde{q}_n = q_n(G\lambda)^n, n = 0, 1, 2, \ldots$.

This expression is the Newtonian potential for an axially symmetric distribution of masses. It is determined with the help of the Newtonian moments

$$N_n = (-1)^n \frac{n!}{(2n+1)!!} \tilde{q}_n M^{n+1},$$

or in the system $G = c = 1$,

$$N_n = (-1)^n \frac{n!}{(2n+1)!!} q_n m^{n+1}. \qquad (1.138)$$

The procedure of calculation the relativistic multipole Geroch–Hansen moments M_n gives the first five terms in the form ($q_0 = 1$)

$$M_0 = N_0 = m, \quad M_1 = N_1 = -\frac{1}{3} q_1 m^2, \quad M_2 = N_2 = \frac{2}{15} q_2 m^3,$$

$$M_3 = -\frac{2}{35} q_3 m^4 + \frac{2}{15} q_1 m^4, \quad M_4 = \frac{8}{315} q_4 m^5 - \frac{4}{105} q_2 m^5 - \frac{2}{21} q_1^2 m^5.$$

$$(1.139)$$

3. Another method of constructing the gravitational multipoles by using the property of a harmonic function is proposed in Ref. [79].

According to it, if the harmonic function ψ is a solution of equation (1.11), then its derivative $\partial\psi/\partial z$ also satisfies this equation.

Thus, one can construct the solution of equation (1.11) in the form

$$\psi = \psi_0 + a_n \widehat{L}_n \psi_0, \tag{1.140}$$

where ψ_0 is an arbitrary solution of equation (1.11), a_n is a constant, and the differential operator $\widehat{L}_n$ is defined as follows:

$$\widehat{L}_n = k_0^n \frac{\partial^n}{\partial z^n} = \left\{ \frac{1}{x^2 - y^2} \left[y(x^2 - 1)\frac{\partial}{\partial x} + x(1 - y^2)\frac{\partial}{\partial y} \right] \right\}^n. \tag{1.141}$$

If one takes the Schwarzschild solution as ψ_0, the function

$$\psi = \frac{1}{2}\ln\left(\frac{x-1}{x+1}\right) + \frac{1}{2}a_n\widehat{L}_n\ln\left(\frac{x-1}{x+1}\right) \tag{1.142}$$

generalizes the Schwarzschild solution for the case of the gravitational field of a mass possessing nth-order multipole moments.

To sum up, one should say that the Schwarzschild solution contains a single parameter m, which characterizes the mass of a source of the gravitational field. Since it is the unique spherically symmetric solution of the vacuum Einstein equations, it has a unique physical interpretation.

If one considers axially symmetric solutions of gravistatics, then construction of gravitational multipoles becomes ambiguous. It means that different solutions can give asymptotically the same Newtonian limit.

References

[1] A. Einstein, *Sitzungsber. Preuss. Akad. Wiss.* 48, 844 (1915).
[2] K. Schwarzschild, *Sitzungsber. Preuss. Akad. Wiss.* 7, 189 (1916).
[3] H. Weyl, *Ann. Phys.* 54, 117 (1917).
[4] I. Chazy, *Bull. Soc. Math. Fr.* 52, 17 (1924).
[5] H. Curzon, *Proc. Math. Soc. Lond.* 23, 477 (1924).
[6] G. Erez, N. Rosen, *Bull. Res. Counc. Isr.* F8, 47 (1959).
[7] V. I. Bashkov, in *Results of Science and Engineering. Algebra. Topology. Geometry*, Vol. 14, VINITI (1976), p. 281 (in Russian).

[8] J. Ehlers, W. Kundt, in L. Witten (ed.), *Gravitation: An Introduction to Current Research*, Wiley, New York (1962).

[9] W. Kinnersley, in G. Shaviv, J. Rosen (eds.), *General Relativity and Gravitation. Proceedings of GR 7, Tel-Aviv, 1974*, Wiley, New York (1975), p. 109.

[10] C. Reina, A. Treves, *Gen. Rel. Grav.* 7, 817 (1976).

[11] D. Kramer, H. Stephani, E. Herlt, M. MacCallum, *Exact Solution of Einstein's Field Equation*, Cambridge University Press, London (1980).

[12] R. Gautreau, J.L. Anderson, *Phys. Lett.* 25A, 291 (1967).

[13] P.G. Bergman, *Introduction to the Theory of Relativity*, Prentice-Hall, Inc., Englewood Cliffs (1942)

[14] A. Einstein, N. Rosen, *Phys. Rev.* 49, 401 (1936).

[15] B. Hoffman, in *Proceeding of the Royaumont Conference*, Centre National de la Recherche Scientifique, Paris (1962).

[16] L. Silberstein, *Phys. Rev.* 49, 268 (1936).

[17] L. Bell, *Gen. Rel. Grav.* 1, 337 (1971).

[18] W.C. Hernandez, *Phys. Rev.* 153, 1359 (1967).

[19] J.J.J. Marek, *Phys. Rev.* 163, 1373 (1967).

[20] J. Stachel, *Phys. Lett.* 27A, 60 (1968).

[21] P. Szekeres, F.H. Morgan, *Commun. Math. Phys.* 32, 313 (1976).

[22] B.H. Voorhees, *Phys. Rev.* D2, 2119 (1970).

[23] F.I. Cooperstock, G.J.G. Junevicus, *Int. J. Theor. Phys.* 9, 59 (1974).

[24] S.M. Scott, P. Szekeres, *Gen. Rel. Grav.* 18, 557 (1986).

[25] P. Szekeres, *Phys. Rev.* 176, 1446 (1968).

[26] A. Banerjee, L.G. Andrade, *Prog. Theor. Phys.* 63, 704 (1980).

[27] S. Chaudhuri, S. Banerji, *Gen. Rel. Grav.* 16, 375 (1984).

[28] G.A. Alekseev, *Sov. Phys. Dokl.* 26, 158 (1981).

[29] C.W. Misner, K.S. Thorne, J.A. Wheeler, *Gravitation*, Freeman, San Francisco (1973).

[30] S. Chandrasekhar, *The Mathematical Theory of Black Holes*, Oxford, Clarendon Press (1983).

[31] W.B. Bonnor, *Gen. Rel. Grav.* 24, 551 (1992).

[32] R. Gautreau, *Nuovo Cim.* 56 B, 49 (1980).

[33] G.D. Birkhoff, *Relativity and Modern Physics*, Harvard University Press, Cambridge (1923).

[34] W. Israel, *Phys. Rev.* 164, 1776 (1967).

[35] T.I. Gutsunaev, Y.G. Ermolaev, in *Problems of Statistical Physics*, RUDN, Moscow (1981).

[36] D.M. Zipoy, *J. Math. Phys.* 7, 1137 (1966).

[37] R.M. Misra, *Phys. Rev.* D2, 410 (1970).

[38] W.B. Bonnor, J.M. Sackfield, *Commun. Math. Phys.*, 8, 338 (1968).

[39] A. Chamorro, R. Gregpry, J.M. Stewart, *Proc. R. Soc. Lond.* A413, 251 (1987).

[40] R.M. Misra, *Proc. Natl. Inst. Sci. India* 27A, 373 (1961).

[41] A.F. Teixeira, *Prog. Theor. Phys.* 60, 163 (1978).

[42] T.I. Gutsunaev, V.A. Chernyaev, *Iz. Vyssh. Uchebn. Zaved. Fiz.* 26, 120 (1982).

[43] R. Bach, H. Weyl, *Math. Zeitschrift.* 13, 134 (1922).

[44] J. Bicak, D. Lynden-Bell, J. Katz, *Phys. Rev.* D47, 4334 (1993).

[45] J. Bicak, D. Lynden-Bell, C. Pichon, *MNRAS* 265, 126 (1993).

[46] J. Bicak, T. Ledvinka, *Phys. Rev. Lett.* 71, 1669 (1993).

[47] C. Pichon, D. Lynden-Bell, *MNRAS* 280, 1007 (1996).

[48] C. Pichon, D. Lynden-Bell, *MNRAS* 282, 1143 (1996).

[49] P.C. Waylen, *Proc. R. Soc. Lond.* A382, 467 (1982).

[50] J.L. Synge, *Relativity: The General Theory*, North-Holland, Amsterdam (1960).

[51] G.E. Tauber, *Canad. J. Phys.* 35, 477 (1957).

[52] W. Dietz, C. Hoenselaers, *Proc. R. Soc. Lond.* A382, 221–229 (1982).

[53] K.S. Das, *Phys. Rev.* D27, 322 (1983).

[54] V.A. Belinsky, V.E. Zakharov, *Sov. Phys. JETP* 48, 985 (1978).

[55] V.A. Belinsky, V.E. Zakharov, *Sov. Phys. JETP* 50, 1 (1979).

[56] S. Chaudhuri, K.C. Das, *Gen. Rel. Grav.* 30(4), 659 (1998).

[57] G.A. Alekseev, V.A. Belinsky, *Sov. Phys. JETP* 78, 1297 (1980).

[58] H. Bondi, *Rev. Mod. Phys.* 29, 23 (1957).

[59] V.P. Kaigorodov, in *Gravitation and Relativity Theory*, Kazan, KGU, Issue 4–5, 3 (1962) (in Russian).

[60] G. Darmois, *Memorial des Sciences Mathematiques*, Fasc. 25, Gauthier-Villars, Paris (1927).

[61] W. Israel, K.A. Khan, *Nuovo Cim.* 33, 331 (1964).

[62] R. Gautreau, R.B. Hoffman, A. Armenti, *Nuovo Cim.* 7B, 71 (1972).

[63] S.P. Gavrilov, in *Gravitation and Relativity Theory*, Kazan, KGU, Issue 8, 30 (1962) (in Russian).

[64] F.P. Esposito, L. Witten, *Phys. Lett.* 58B,357 (1975).

[65] D. Papadopoulos, B.W. Stewart, L. Witten, *Phys. Rev.* D24, 320 (1981).

[66] R.M. Kerns, W.J. Wild, *Phys. Rev.* D26, 3726 (1982).

[67] T. I. Gutsunaev, *Izv. Vyssh. Ucheb. Zaved. Fiz.* 11, 145 (1976) (in Russian).

[68] R.J. Geroch, *J. Math. Phys.* 11, 2560 (1970).

[69] R.O. Hansen, *J. Math. Phys.* 15, 46 (1974).

[70] L.D. Landau, E.M. Lifshitz, *The Classical Theory of Fields*, Pergamon Press, Oxford (1983).

[71] S. Chandrasekhar, *The Mathematical Theory of Black Holes*, Clarendon Press, Oxford (1983).

[72] C. Hoenselaers, in H. Sato, T. Nakamura (eds.), *Gravitational Collapse and Relativity. Proceedings of 14th Yamadda Conference*, Kyoto, Japan, World Scientific, Singapore (1986), p. 176.

[73] S. Chandrasekhar, *Proc. R. Soc. Lond.* A376, 333 (1981).

[74] R. Beig, W. Simon, *Proc. R. Soc. Lond.* A376, 333 (1981).

[75] R. Beig, W. Simon, *J. Math. Phys.* 24, 1163 (1983).

[76] K.S. Thorne, *Rev. Mod. Phys.* 52, 299 (1980).

[77] J. Ehlers, in J. Nitsch, J. Pfarr, E.W. Stachov (eds.), *Grundlagenprobleme der Modernen Physik*, BI-Verlag, Mannheim (1981).

[78] Y. Gursel, *Gen. Rel. Grav.* 12, 1003 (1983).

[79] H. Quevedo, *Gen. Rel. Grav.* 19, 1013 (1987).

[80] H. Quevedo, *Phys. Rev.* D39, 2904 (1989).

[81] A.G. Doroshkevich, J.B. Zeldovich, I.D. Novikov, *Sov. Phys. JETP* 49, 170 (1965).

[82] T.I. Gutsunaev, V.S. Manko, *Gen. Rel. Grav.* 17, 1025 (1985).

[83] A.A. Shaideman, Generation of vacuum axially symmetric solutions to the equations of General Relativity using stationary Euclidons, PhD thesis, Moscow, 2005 (in Russian).

Chapter 2

Stationary Solutions of the Vacuum Einstein's Equations

2.1 Introduction

The first exact solutions of the axially symmetric stationary vacuum equations were presented by Lewis [1] and Van Stockum [2]. Later, Papapetrou [3] obtained a canonical form of the line element, and Ernst [4] formulated the stationary vacuum problem in the form of a single equation for a complex function. An outstanding result was obtained in 1963 by Kerr [5], whose solution possibly describes the exterior gravitational field of a rotating source. A family of a asymptotically flat solutions of the stationary vacuum equations was found by Tomimatsu and Sato [6, 7]. Later, a survey of known solutions was given by Kinnersley [8]. Possible generalizations of the Tomimatsu–Sato metric were found by Yamazaki [9, 10]. Hori [11] and Cosgrove [12, 13], and asymptotically not-flat solutions of the Ernst problem were obtained in Refs. [14–16]. The Ernst equation [4] makes it possible to extract the solutions by the method of separation of variables [17, 18]. In a series of papers [19–21], a non-canonical form of the stationary metrics was used to obtain new solutions of the Einstein equations. An approximate method for constructing solutions depending on two harmonic functions was proposed in Refs. [22, 23]. Ernst [24] dreamt of construction of such a kind of solutions but exact ones.

2.2   Basic Equations

As was shown by Papapetrou [3], the line element, describing a stationary axially symmetric gravitational field without loss of generality can be present in the following canonical form:

$$ds^2 = f^{-1}[e^{2\gamma}(d\rho^2 + dz^2) + \rho^2 d\varphi^2] - f(dt - \omega d\varphi)^2, \qquad (2.1)$$

where ρ, φ, z and t are canonical Weyl coordinates and time, respectively, $f(\rho, z)$, $\gamma(\rho, z)$ and $\omega(\rho, z)$ are unknown functions to be determined from the field equations.

The vacuum Einstein equations are given by

$$R_{ik} = 0,$$

where R_{ik} is the Ricci tensor.

Therefore, we obtain all the Einstein equations in the case of a stationary axially symmetric gravitational field outside the sources:

$$f\Delta f = (\vec{\nabla} f)^2 - f^4 \rho^{-2} (\vec{\nabla}\omega)^2, \qquad (2.2)$$

$$\vec{\nabla}(f^2 \rho^{-2} \vec{\nabla}\omega) = 0, \qquad (2.3)$$

$$\frac{\partial\gamma}{\partial\rho} = \frac{1}{4}\rho f^{-2}\left[\left(\frac{\partial f}{\partial\rho}\right)^2 - \left(\frac{\partial f}{\partial z}\right)^2 - f^4\rho^{-2}\left(\left(\frac{\partial\omega}{\partial\rho}\right)^2 - \left(\frac{\partial\omega}{\partial z}\right)^2\right)\right],$$

$$(2.4)$$

$$\frac{\partial\gamma}{\partial z} = \frac{1}{2}\rho f^{-2}\left(\frac{\partial f}{\partial\rho}\frac{\partial f}{\partial z} - f^4\rho^{-2}\frac{\partial\omega}{\partial\rho}\frac{\partial\omega}{\partial z}\right). \qquad (2.5)$$

The operators $\vec{\nabla}$ and Δ are defined by the formulae

$$\vec{\nabla} \equiv \vec{\rho}_0 \frac{\partial}{\partial\rho} + \vec{z}_0 \frac{\partial}{\partial z},$$

$$\Delta \equiv \vec{\nabla}^2 \equiv \frac{\partial^2}{\partial\rho^2} + \frac{1}{\rho}\frac{\partial}{\partial\rho} + \frac{\partial^2}{\partial z^2}$$

($\vec{\rho}_0$ and $\vec{z}_0$ being unit vectors), i.e., they are similar to the ordinary Laplacian and gradient operators for flat space expressed in cylindrical coordinates provided that there is no angular coordinate dependence.

It should be noted that the integrability condition of equations (2.4), (2.5) for determining the function $\gamma(\rho, z)$ is systems (2.2), (2.3) which does not contain $\gamma(\rho, z)$. Hence, the problem of obtaining exact axially symmetric solutions of the Einstein equations outside the sources reduces to integration of the differential equations (2.2), (2.3).

In the prolate ellipsoidal coordinates (x, y), the operators Δ and $\vec{\nabla}$ take the form

$$\Delta \equiv \frac{1}{k_0^2(x^2 - y^2)}\left(\frac{\partial}{\partial x}\left[(x^2 - 1)\frac{\partial}{\partial x}\right] + \frac{\partial}{\partial y}\left[(1 - y^2)\frac{\partial}{\partial y}\right]\right),$$

$$\vec{\nabla} \equiv \frac{1}{k_0(x^2 - y^2)^{1/2}}\left(\vec{x}_0\sqrt{x^2 - 1}\frac{\partial}{\partial x} + \vec{y}_0\sqrt{1 - y^2}\frac{\partial}{\partial y}\right),$$

$\vec{x}_0$ and $\vec{y}_0$ being unit vectors, and equations (2.4), (2.5) take the form

$$\frac{\partial\gamma}{\partial x} = \frac{1 - y^2}{4f^2(x^2 - y^2)}\left\{x(x^2 - 1)\left[\left(\frac{\partial f}{\partial x}\right)^2 + \left(\frac{\partial \Phi}{\partial x}\right)^2\right] - x(1 - y^2)\right.$$
$$\left. \times \left[\left(\frac{\partial f}{\partial y}\right)^2 + \left(\frac{\partial \Phi}{\partial y}\right)^2\right] - 2y(x^2 - 1)\left(\frac{\partial f}{\partial x}\frac{\partial f}{\partial y} + \frac{\partial \Phi}{\partial x}\frac{\partial \Phi}{\partial y}\right)\right\},$$
$$\tag{2.6}$$

$$\frac{\partial\gamma}{\partial y} = \frac{x^2 - 1}{4f^2(x^2 - y^2)}\left\{y(x^2 - 1)\left[\left(\frac{\partial f}{\partial x}\right)^2 + \left(\frac{\partial \Phi}{\partial x}\right)^2\right] - y(1 - y^2)\right.$$
$$\left. \times \left[\left(\frac{\partial f}{\partial y}\right)^2 + \left(\frac{\partial \Phi}{\partial y}\right)^2\right] + 2x(1 - y^2)\left(\frac{\partial f}{\partial x}\frac{\partial f}{\partial y} + \frac{\partial \Phi}{\partial x}\frac{\partial \Phi}{\partial y}\right)\right\}.$$
$$\tag{2.7}$$

The line element (2.1) in this case can be rewritten as

$$ds^2 = k_0^2 f^{-1}\left[e^{2\gamma}(x^2 - y^2)\left(\frac{dx^2}{x^2 - 1} + \frac{dy^2}{1 - y^2}\right)\right.$$
$$\left. + (x^2 - 1)(1 - y^2)d\varphi^2\right] - f(dt - \omega d\varphi)^2, \tag{2.8}$$

where the metric coefficients f, γ and ω are functions of x and y only.

Accordingly, in this case, the equation

$$\Delta\psi \equiv \frac{\partial^2\psi}{\partial\rho^2} + \frac{1}{\rho}\frac{\partial\psi}{\partial\rho} + \frac{\partial^2\psi}{\partial z^2} = 0 \tag{2.9}$$

becomes

$$\frac{\partial}{\partial x}\left[(x^2-1)\frac{\partial\psi}{\partial x}\right] + \frac{\partial}{\partial y}\left[(1-y^2)\frac{\partial\psi}{\partial y}\right] = 0, \tag{2.10}$$

and the equation

$$\Delta'\chi \equiv \frac{\partial^2\chi}{\partial\rho^2} - \frac{1}{\rho}\frac{\partial\chi}{\partial\rho} + \frac{\partial^2\chi}{\partial z^2} = 0 \tag{2.11}$$

can be rewritten as

$$(x^2-1)\frac{\partial^2\chi}{\partial x^2} + (1-y^2)\frac{\partial^2\chi}{\partial y^2} = 0. \tag{2.12}$$

There exist two possibilities of writing down equations (2.2), (2.3) in a coordinate-free form:

1. If we switch from the rotation potential ω to the new potential Φ by the formulae

$$\frac{\partial\omega}{\partial\rho} = \frac{\rho}{f^2}\frac{\partial\Phi}{\partial z}, \qquad \frac{\partial\omega}{\partial z} = -\frac{\rho}{f^2}\frac{\partial\Phi}{\partial\rho}, \tag{2.13}$$

or, in the prolate ellipsoidal coordinates,

$$\frac{\partial\omega}{\partial x} = \frac{k_0(1-y^2)}{f^2}\frac{\partial\Phi}{\partial y}, \qquad \frac{\partial\omega}{\partial y} = -\frac{k_0(x^2-1)}{f^2}\frac{\partial\Phi}{\partial x}, \tag{2.14}$$

we obtain the following field equations:

$$f\Delta f = (\vec{\nabla}f)^2 - (\vec{\nabla}\Phi)^2, \qquad \vec{\nabla}(f^{-2}\vec{\nabla}\Phi) = 0. \tag{2.15}$$

The set of two second-order differential equations can be reduced to four first-order differential equations.

Let us introduce the notations

$$f^{-1}\vec{\nabla}f = \vec{A}, \quad f^{-1}\vec{\nabla}\Phi = \vec{B}. \tag{2.16}$$

In this case, we obtain the following set of four first-order differential equations for $\vec{A}$ and $\vec{B}$:

$$\begin{aligned}
\operatorname{div}\vec{A} &= -\vec{B}^2, \quad \operatorname{div}\vec{B} = (\vec{A}\cdot\vec{B}), \\
\operatorname{rot}\vec{A} &= 0, \quad \operatorname{rot}\vec{B} = -[\vec{A}\times\vec{B}],
\end{aligned} \tag{2.17}$$

This form of the field equations was developed in Refs. [25, 26]. In the Weyl canonical coordinates, we can rewrite (2.16) as

$$\begin{aligned}
A_1 &= \frac{1}{f}\frac{\partial f}{\partial\rho}, \quad A_2 = \frac{1}{f}\frac{\partial f}{\partial z}, \\
B_1 &= \frac{1}{f}\frac{\partial\Phi}{\partial\rho}, \quad B_2 = \frac{1}{f}\frac{\partial\Phi}{\partial z}.
\end{aligned} \tag{2.18}$$

In this case, using (2.17) and (2.18), we obtain

$$\begin{aligned}
\frac{1}{\rho}\frac{\partial}{\partial\rho}(\rho A_1) + \frac{\partial A_2}{\partial z} &= -(B_1^2 + B_2^2), \\
\frac{1}{\rho}\frac{\partial}{\partial\rho}(\rho B_1) + \frac{\partial B_2}{\partial z} &= A_1 B_1 + A_2 B_2, \\
\frac{\partial A_1}{\partial z} - \frac{\partial A_2}{\partial\rho} = 0, \quad \frac{\partial B_1}{\partial z} - \frac{\partial B_2}{\partial\rho} &= A_2 B_1 - A_1 B_2.
\end{aligned} \tag{2.19}$$

Let us consider an analogue of the system (2.19) in the ellipsoidal coordinates (x, y).

Introducing the functions

$$\begin{aligned}
\alpha_1 &= \frac{1}{f}\frac{\partial f}{\partial x}, \quad \alpha_2 = \frac{1}{f}\frac{\partial f}{\partial y}, \\
\beta_1 &= \frac{1}{f}\frac{\partial\Phi}{\partial x}, \quad \beta_2 = \frac{1}{f}\frac{\partial\Phi}{\partial y},
\end{aligned} \tag{2.20}$$

we have from equations (2.19) the following equivalent set of four first-order differential equations:

$$\frac{\partial}{\partial x}[(x^2-1)\alpha_1] + \frac{\partial}{\partial y}[(1-y^2)\alpha_2] = -(x^2-1)\beta_1^2 - (1-y^2)\beta_2^2,$$

$$\frac{\partial}{\partial x}[(x^2-1)\beta_1] + \frac{\partial}{\partial y}[(1-y^2)\beta_2] = (x^2-1)\alpha_1\beta_1 + (1-y^2)\alpha_2\beta_2,$$

$$\frac{\partial\alpha_2}{\partial x} - \frac{\partial\alpha_1}{\partial y} = 0, \quad \frac{\partial\beta_2}{\partial x} - \frac{\partial\beta_1}{\partial y} = \alpha_2\beta_1 - \alpha_1\beta_2. \tag{2.21}$$

Introducing the Ernst complex potential

$$\varepsilon = f + i\Phi \tag{2.22}$$

gives us, from (2.15), the equation

$$(\varepsilon + \varepsilon^*)\Delta\varepsilon = 2(\vec{\nabla}\varepsilon)^2, \tag{2.23}$$

where ε^* is the complex conjugate of ε.

A transformation of (2.23)

$$\varepsilon = (\xi - 1)(\xi + 1)^{-1} \tag{2.24}$$

leads to one more forms of the equations for a stationary axially symmetric gravitational field:

$$(\xi\xi^* - 1)\Delta\xi = 2\xi^*(\vec{\nabla}\xi)^2. \tag{2.25}$$

For equation (2.23), we can prove the following theorem:

Theorem 1. *If $\tilde{\varepsilon}$ is a solution of equation, then the functions*

$$\varepsilon = \frac{iA_0 + B_0\tilde{\varepsilon}}{C_0 + iD_0\tilde{\varepsilon}}, \quad \varepsilon^* = \frac{-iA_0 + B_0\tilde{\varepsilon}^*}{C_0 - iD_0\tilde{\varepsilon}^*}, \tag{2.26}$$

where A_0, B_0, C_0 and D_0 are arbitrary real constants subject to the constraint $A_0D_0 + B_0C_0 \neq 0$, also satisfy equation (2.23).

Its proof can be obtained immediately by substitution of into the corresponding equation in (2.23).

From the equations (2.26), with the help of (2.22), one can obtain

$$f = \frac{(A_0 D_0 + B_0 C_0)\tilde{f}}{(C_0 - D_0\tilde{\Phi})^2 + D_0^2 \tilde{f}^2},$$

$$\Phi = \frac{A_0 C_0 + (B_0 C_0 - A_0 D_0)\tilde{\Phi} - B_0 D_0(\tilde{f}^2 + \tilde{\Phi}^2)}{(C_0 - D_0\tilde{\Phi})^2 + D_0^2 \tilde{f}^2} + E_0, \tag{2.27}$$

where $\tilde{f}$ and $\tilde{\Phi}$ satisfy the equations (2.15). It should be noted that if we let in $A_0 = E_0 = 0$, $B_0 = C_0 = 1$, we obtain the Ehlers transformation ([34], see also Ref. [8])

$$f = \frac{\tilde{f}}{(1 - D_0\tilde{\Phi})^2 + D_0^2 \tilde{f}^2}, \quad \Phi = \frac{\tilde{\Phi} - D_0(\tilde{f}^2 + \tilde{\Phi}^2)}{(1 - D_0\tilde{\Phi})^2 + D_0^2 \tilde{f}^2}. \tag{2.28}$$

The inverse transformation

$$\varepsilon = \frac{1}{\tilde{\varepsilon}}, \quad \varepsilon^* = \frac{1}{\tilde{\varepsilon}^*} \tag{2.29}$$

is recognized to be a special case of (2.26) corresponding to the choice $B_0 = C_0 = E_0 = 0$, $A_0 = D_0$. So, (2.27) take the form

$$f = \frac{\tilde{f}}{\tilde{f}^2 + \tilde{\Phi}^2}, \quad \Phi = \frac{\tilde{\Phi}}{\tilde{f}^2 + \tilde{\Phi}^2}. \tag{2.30}$$

It can be pointed out that the inverse transformation leaves the metric function γ unchanged, i.e., $\gamma = \tilde{\gamma}$.

As an another particular case of (2.26), the constant phase transformation

$$\xi = e^{i\alpha_0}\tilde{\xi}, \tag{2.31}$$

where α_0 is real constant and the function $\tilde{\xi}$ is introduced according (2.24), gives us a possibility of obtaining new solutions of the Ernst equation (2.25) from old ones.

The transformation

$$\varphi \to C_1\varphi + C_2 t, \quad t \to C_3\varphi + C_4 t, \tag{2.32}$$

where C_1, C_2, C_3 and C_4 are real constants, preserves the characteristic form of the Papapetrou line element (2.1). In this connection, we can proof the following theorem.

2. The substitution $f = \rho/F$ in equations gives us the following equations:

$$F\Delta F = (\vec{\nabla} F)^2 + (\vec{\nabla}\omega)^2, \quad \vec{\nabla}(F^{-2}\vec{\nabla}\omega) = 0. \qquad (2.33)$$

The set of two second-order differential equations can also be reduced to a set of four first-order differential equations.

If we introduce the notations

$$F^{-1}\vec{\nabla} F = \vec{A}, \quad F^{-1}\vec{\nabla}\omega = \vec{B}, \qquad (2.34)$$

we obtain (2.17).

Introducing the functions

$$\varepsilon_1 = F + \omega \quad \varepsilon_2 = F - \omega, \qquad (2.35)$$

we rewrite equations (2.33) in the symmetric form

$$(\varepsilon_1 + \varepsilon_2)\Delta\varepsilon_1 = 2(\vec{\nabla}\varepsilon_1)^2,$$
$$(\varepsilon_1 + \varepsilon_2)\Delta\varepsilon_2 = 2(\vec{\nabla}\varepsilon_2)^2. \qquad (2.36)$$

For the field equations (2.36) and (2.33) one can prove a theorem similar to (2.26) and (2.27). So, for the equations (2.36) we have the following theorem.

Theorem 2. *If $\tilde{\varepsilon}_1$ and $\tilde{\varepsilon}_2$ are solution of equations, then the functions*

$$\varepsilon_1 = \frac{A_0 + B_0\tilde{\varepsilon}_1}{C_0 + D_0\tilde{\varepsilon}_1}, \quad \varepsilon_2 = \frac{-A_0 + B_0\tilde{\varepsilon}_2}{C_0 - D_0\tilde{\varepsilon}_2}, \qquad (2.37)$$

where A_0, B_0, C_0 and D_0 are arbitrary real constants subject to the constraint $B_0C_0 - A_0D_0 \neq 0$, also satisfy equations.

From (2.37) with the help of (2.35), we obtain

$$F = \frac{(B_0 C_0 - A_0 D_0)\tilde{F}}{(C_0 + D_0\tilde{\omega})^2 - D_0^2 \tilde{F}^2},$$

$$\omega = \frac{A_0 C_0 + (A_0 D_0 + B_0 C_0)\tilde{\omega} - B_0 D_0(\tilde{F}^2 - \tilde{\omega}^2)}{(C_0 + D_0\tilde{\omega})^2 - D_0^2 \tilde{F}^2},$$

(2.38)

where $\tilde{F}$ and $\tilde{\omega}$ satisfy equation (2.23).

References [27–30] are devoted to the symmetric of vacuum stationary axially symmetric Einstein equations.

One more possible formulation of the axially symmetric equations was proposed in 1978 by Cosgrove [35, 36]. Rather than examining the system of two second-order partial differential equations for the function f and Φ, Cosgrove preferred to analyze one fourth-order partial differential equation for the metric function γ.

Finally, it should be noted that the procedure for calculating the relativistic multipole moments was given by Geroch [37, 38] and Hansen [39].

A simplified procedure for calculating the Geroch and Hansen multipole moments was proposed by Hoenselaers [40], who found a relation between the Ernst potential of an arbitrary axially symmetric stationary vacuum metric and the corresponding Geroch–Hansen multipole moments.

The relativistic and coordinate-invariant definitions of multipole moments were also proposed in Refs. [41, 42]. Although one is led to these definitions by different mathematical approaches, it can be shown that all of them are physically equivalent [43, 44].

2.3 Lewis' Class of Solutions: One-Stationary Euclidon Solution

In 1932, Lewis [1] obtained the following class of exact solutions of equations ((2.13), (2.15)):

$$f_L = \frac{\rho}{C_1}\frac{\sinh(\psi + U_0)}{\cosh U_0},$$

$$\Phi_L = \frac{1}{C_1 \cosh U_0} \frac{\partial \chi}{\partial z} + C_2,$$

$$\omega_L = C_1 \frac{\cosh \psi}{\sinh(\psi + U_0)} + C_3, \tag{2.39}$$

where C_1, C_2, C_3 and U_0 are arbitrary constants and $\psi(\rho, z) = (1/\rho)(\partial \chi/\partial \rho)$ and $\chi(\rho, z)$ are functions satisfying the linear equations (2.9), (2.11).

Note that Weyl's static solution $f_W = (\rho/C_1)e^{-\psi}$, $\Phi_W = C_2$ and $\omega_W = C_3$ (where C_2 and C_3 are usually set equal to zero) can be obtained from (2.39) in the limit $U_0 \to \infty$.

To obtain the basic classes of stationary vacuum solutions in a more or less unified way, we use the representation of equations (2.15) in the form of four first-order differential equations (2.19) for the functions (2.18), which we further subject to certain algebraic dependences.

The class of solutions (2.39) can be obtained by making the functions (2.18) satisfy the condition

$$A_1 B_1 + A_2 B_2 = \frac{B_1}{\rho}. \tag{2.40}$$

Then, one can introduce the new function $M(\rho, z)$ such that

$$B_1 = M A_2, \quad B_2 = -M \left(A_1 - \frac{1}{\rho} \right). \tag{2.41}$$

Then, from (2.19), we have the equation

$$M(M^2 - 1)\Delta M = (2M^2 - 1)(\vec{\nabla} M)^2. \tag{2.42}$$

If the function $M(\rho, z)$ is known, one can easily find the functions F and Φ from equations (2.18) with help of the relations

$$A_1 = \frac{1}{M(M^2 - 1)} \frac{\partial M}{\partial \rho} + \frac{1}{\rho}, \quad A_2 = \frac{1}{M(M^2 - 1)} \frac{\partial M}{\partial z},$$

$$B_1 = \frac{1}{M^2 - 1} \frac{\partial M}{\partial z}, \quad B_2 = \frac{1}{1 - M^2} \frac{\partial M}{\partial \rho}. \tag{2.43}$$

1. Let us consider the most important cases of the class of solutions under study.

1-I. $M^2 > 1$. In this case, equation (2.42), by the substitution

$$M = \frac{1}{\cos(\psi + U_0)},\tag{2.44}$$

reduces to the linear equation (2.9). In this case, we obtain the solutions

$$f = \frac{\rho}{\cos U_0}\sin(\psi + U_0),\quad \Phi = \frac{1}{\cos U_0}\frac{\partial \chi}{\partial z},$$

$$\omega = \frac{\cos\psi}{\sin(\psi + U_0)},\tag{2.45}$$

$$\gamma = \frac{1}{2}\ln\left(\sin\psi + \frac{\cos\psi}{\cos U_0}\right) + \frac{1}{4}\ln\rho + \gamma',$$

where the functions ψ and χ satisfy equations (2.9) and (2.11), respectively, and are related to each other by $\psi = (1/\rho)(\partial\chi/\partial\rho)$. The function γ' can be obtained from equations.

The function γ' is defined by

$$\frac{\partial\gamma'}{\partial\rho} = \frac{\rho}{4}\left[\left(\frac{\partial\psi}{\partial\rho}\right)^2 - \left(\frac{\partial\psi}{\partial z}\right)^2\right],\quad \frac{\partial\gamma'}{\partial z} = \frac{\rho}{2}\frac{\partial\psi}{\partial\rho}\cdot\frac{\partial\psi}{\partial z}.\tag{2.46}$$

It should be noted that solution (2.45) can be transformed to the static vacuum Weyl class by a complex linear transformation of the coordinates φ and t.

1-II. $M^2 = 1$.

The solution in this case has the form

$$f = \frac{\partial\chi}{\partial\rho},\quad \Phi = \frac{\partial\chi}{\partial z},$$

$$\omega = -\rho\left(\frac{\partial\chi}{\partial\rho}\right)^{-1},\quad \gamma = \frac{1}{2}\ln\frac{\partial\chi}{\partial\rho} - \frac{1}{4}\ln\rho,\tag{2.47}$$

$$\Delta'\chi = 0,$$

and it determines the van Stockum metric [2].

1-III. $M^2 < 1$. In this case,

$$M = \cosh^{-1}(\psi + U_0) \quad (U_0 = \text{const}), \tag{2.48}$$

and we obtain (2.39).

The metric function γ is in this case

$$\gamma = \gamma_L = \frac{1}{2} \ln \left(\sinh \psi + \frac{\cosh \psi}{\cosh U_0} \right) + \frac{1}{4} \ln \rho + \gamma'. \tag{2.49}$$

The function γ' can be obtained from (2.46).

Equation (2.39) determine the static vacuum Weyl class because the function ω can be eliminated by a linear transformation

$$\varphi \to C_1^0 \varphi + C_2^0 t, \quad t \to C_3^0 \varphi + C_4^0 t,$$

where C_1^0, C_2^0, C_3^0 and C_4^0 are real constants.

(a) If one takes the expression

$$\chi = \chi_1 = \int \sqrt{\rho^2 + (z - z_{1]})^2} dz,$$

then

$$\psi = \psi_1 = \frac{1}{\rho} \frac{\partial \chi_1}{\partial \rho} = -\ln \rho + \ln[(z - z_1 + \sqrt{\rho^2 + (z - z_1)^2})],$$

where z_1 is a displacement constant.

The solution in this particular case has the form

$$f = f_1 = (z - z_1) + \sqrt{\rho^2 + (z - z_1)^2} \cdot \tanh U_0,$$

$$\Phi = \Phi_1 = \frac{\sqrt{\rho^2 + (z - z_1)^2}}{\cosh U_0}, \quad \omega = \omega_1 = \frac{\sqrt{\rho^2 + (z - z_1)^2}}{\cosh U_0} \cdot \frac{1}{f_1},$$

$$\gamma = \gamma_1 = \frac{1}{2} \ln f_1 - \frac{1}{2} \ln \sqrt{\rho^2 + (z - z_1)^2}. \tag{2.50}$$

The line element then takes the form

$$ds^2 = \frac{d\rho^2 + dz^2}{\sqrt{\rho^2 + (z - z_1)^2}} + \frac{\rho^2}{f_1} d\varphi^2$$

$$- f_1 \left[dt - \frac{\sqrt{\rho^2 + (z - z_1)^2}}{\cosh U_0} \cdot \frac{1}{f_1} d\varphi \right]^2. \tag{2.51}$$

It can be checked that a direct substitution of the solution (2.50) nullifies all components of the Riemann tensor, which corresponds to flat space.

So, we call the solution a stationary Euclidon.

(b) If we choose the function χ in the form

$$\chi = \chi_2 = \frac{z^2}{2} - \frac{\rho^2}{4}(2\ln\rho - 1),$$

then we obtain the solutions

$$f = f_2 = W(\rho), \quad \Phi = \Phi_2 = \frac{z}{\cosh U_0},$$

$$\omega = \omega_2 = \frac{1+\rho^2}{\rho\cosh U_0}\frac{1}{W(\rho)}, \tag{2.52}$$

$$\gamma = \gamma_2 = \frac{\ln W(\rho)}{2}, \quad W(\rho) = \frac{1+\tanh U_0}{2} - \frac{1-\tanh U_0}{2}\cdot\rho^2,$$

which is a special case of the general stationary cylindrically symmetric solution [45], i.e., the Kasner metric ([46], see also Refs. [47–51]). The line element in this case takes the form

$$ds^2 = d\rho^2 + dz^2 + \frac{\rho^2}{W(\rho)}d\varphi^2 - W(\rho)\left[dt - \frac{1+\rho^2}{\rho\cosh U_0}\frac{1}{W(\rho)}d\varphi\right]^2, \tag{2.53}$$

which can be obtained from the Minkowski metric by transformation

$$\varphi' = \frac{1+\tanh U_0}{2}\varphi + \frac{1-\tanh U_0}{2}t, \quad t' = \frac{1+\tanh U_0}{2}t. \tag{2.54}$$

Therefore, we call the solution a stationary Euclidon of second kind.

2. Let us finally consider simple methods of solving equations.

2-I. We rewrite equations (2.33) in the form

$$\vec{\nabla}\left[\frac{\vec{\nabla}(F^2 - \omega^2)}{F^2}\right] = 0, \quad \vec{\nabla}\left(\frac{\vec{\nabla}\omega}{F^2}\right) = 0 \tag{2.55}$$

and choose the solution of equations in the form $F = F(\psi)$ and $\omega = \omega(\psi)$, where ψ is an arbitrary solution of the equation $\Delta\psi = 0$.

In this case, integrating the set of ordinary differential equations

$$\frac{d}{d\psi}\left[\frac{1}{F^2}\frac{d}{d\psi}(F^2 - \omega^2)\right] = 0, \qquad \frac{d}{d\psi}\left(\frac{1}{F^2}\frac{d\omega}{d\psi}\right) = 0, \qquad (2.56)$$

we find

$$\frac{d}{d\psi}(F^2 - \omega^2) = a_0 F^2, \qquad \frac{d\omega}{d\psi} = b_0 F^2, \qquad (2.57)$$

$$a_0 = \text{const.}, \qquad b_0 = \text{const.}$$

The integration (2.57), as well as the following choice and renaming of the constants,

$$b_0 = -\frac{1}{C_1\cosh U_0}, \qquad a_0 = \frac{2C_3}{C_1\cosh U_0} - 2\tanh U_0,$$

brings the solution to its final form

$$f = \rho\frac{\sinh(\psi + U_0)}{C_1\cosh U_0}, \qquad \omega = \frac{C_1\cosh\psi}{\sinh(\psi + U_0)} + C_3,$$

i.e., to the Lewis class of solutions.

2-II. The Lewis class of solutions can be easily found with the aid of the symmetry transformation.

In a special case, setting in (2.38),

$$\tilde{F} = e^{-\psi} \quad (\Delta\psi = 0, U_0 = \text{ const.}),$$

$$\tilde{\omega} = 0, \quad A_0 = -D_0 = e^{-U_0}, \quad B_0 = C_0 = 1,$$

we obtain

$$F = \frac{\cosh U_0}{\sinh(\psi + U_0)}, \qquad \omega = \frac{\cosh\psi}{\sinh(\psi + U_0)}.$$

Recalling that $F = \rho/f$, we arrive at the Lewis class of solutions.

2.4   Physical Interpretation of One-Stationary Euclidon Solution

In this section, the physical content of stationary Euclidon solutions to Einstein's equations is analyzed. These solutions serve as "building blocks" of the theory, which allows for the construction of almost all known solutions to the vacuum stationary axially symmetric Einstein equations, including such important ones as the Kerr solution.

One of the solutions to the field equations has the form

$$f = f_E = \frac{z - z_1 + \sqrt{\rho^2 + (z - z_1)^2}\,\tanh V_0}{C_1},$$

$$\omega = \omega_E = \frac{\sqrt{\rho^2 + (z - z_1)^2}}{f_E} \cdot \frac{1}{\cosh V_0} + C_3, \tag{2.58}$$

$$\gamma = \gamma_E = \frac{1}{2}\ln \frac{z - z_1 + \sqrt{\rho^2 + (z - z_1)^2}\,\tanh V_0}{\sqrt{\rho^2 + (z - z_1)^2}} + \gamma_0,$$

where $z_1, C_1, C_3, V_0, \gamma_0$ are constant values. The solution is the so-called stationary Euclidon solution.

According to this solution, the corresponding metric is

$$ds_E^2 = C_1 e^{2\gamma_0} \frac{d\rho^2 + dz^2}{\sqrt{\rho^2 + (z - z_1)^2}} + \frac{\rho^2 d\varphi^2}{f_E}$$

$$- f_E \left[cdt - \left(\frac{\sqrt{\rho^2 + (z - z_1)^2}}{f_E \cdot \cosh V_0} + C_3 \right) d\varphi \right]^2. \tag{2.59}$$

It is easy to see that the transformation [133]

$$2C_1 e^{2\gamma_0} \rho = \rho' \sqrt{(z' - z_1')^2 - c^2 t'^2},$$

$$e^{-\gamma_0}\varphi = \varphi'\sqrt{1 + \tanh V_0} - \frac{\sqrt{1 - \tanh V_0}}{2}\ln\frac{z' - z_1' + ct'}{z' - z_1' - ct'},$$

$$4C_1 e^{2\gamma_0}(z - z_1) = (z' - z_1')^2 - c^2 t'^2 - \rho'^2,$$

$$\frac{\sqrt{1 + \tanh V_0}}{2}\frac{e^{-\gamma_0}ct}{C_1}$$

$$= C_4\frac{1 + \tanh V_0}{2C_1}\varphi' + \left(1 - \frac{C_4}{2C_1\cosh V_0}\right)\frac{1}{2}\ln\frac{z' - z_1' + ct'}{z' - z_1' - ct'}, \tag{2.60}$$

where

$$C_4 = C_3 + C_1 \frac{\sqrt{1 - \tanh V_0}}{\sqrt{1 + \tanh V_0}},$$

transforms the metric into the Minkowski metric

$$ds_M^2 = d\rho'^2 + \rho'^2 d\varphi'^2 + dz'^2 - c^2 dt'^2. \tag{2.61}$$

Thus, the metric of the stationary Euclidon solution describes flat euclidean spacetime.

The formulas for the inverse transformation are as follows:

$$\rho' = e^{\gamma_0} \sqrt{2C_1 \mu_-},$$

$$\varphi' = \frac{e^{-\gamma_0} \varphi}{\sqrt{1 + \tanh V_0}} + \frac{\sqrt{1 - \tanh V_0}}{2} \cdot \frac{e^{-\gamma_0}}{C_1}(ct - C_4\varphi),$$

$$z' - z_1' = \varepsilon e^{\gamma_0} \sqrt{2C_1 \mu_+} \cosh\left[\frac{\sqrt{1 + \tanh V_0}}{2} \frac{e^{-\gamma_0}}{C_1}(ct - C_4\varphi)\right],$$

$$ct' = e^{\gamma_0} \sqrt{2C_1 \mu_+} \sinh\left[\frac{\sqrt{1 + \tanh V_0}}{2} \frac{e^{-\gamma_0}}{C_1}(ct - C_4\varphi)\right],$$

$$\tag{2.62}$$

where

$$\mu_\pm = \sqrt{\rho^2 + (z - z_1)^2} \pm (z - z_1), \quad \varepsilon = \pm 1.$$

Returning to the solution, it can be shown through straightforward but cumbersome calculations that this solution reduces all components of the Riemann curvature tensor R_{iklm} to zero. Therefore, the metric remains within the framework of special relativity (SR).

This fact is the reason for the name "Euclidon."

All of the above indicates that the Euclidon solution is associated with a relativistic reference frame and has a clear physical interpretation.

From now on, we refer to the primed reference frame $(\rho', \varphi', z', t')$ as the stationary or inertial reference frame (IRF) and (ρ, φ, z, t) as the non-inertial reference frame (NIRF). The transformation between the two is carried out using specific formulas.

Now, let's consider a point at rest in the non-inertial system (ρ, φ, z, t)

$$\rho = \rho_0, \quad z = z_0, \quad \varphi = \varphi_0' e^{\gamma_0} \sqrt{1 + \tanh V_0}.$$

For it,

$$\mu_\pm^0 = \sqrt{\rho_0^2 + (z_0 - z_1)^2} \pm (z_0 - z_1). \tag{2.63}$$

We also introduce the notations

$$z' = -\frac{c^2}{\tilde{a}_0}, \quad \tilde{a}_0 = \varepsilon \frac{c^2}{e^{\gamma_0} \sqrt{2C_1 \mu_+^0}}, \tag{2.64}$$

$$\Omega_0 = \frac{\tilde{a}_0}{c} \sqrt{\frac{1 - \tanh V_0}{1 + \tanh V_0}}. \tag{2.65}$$

From the perspective of the inertial reference frame (IRF), this point is in motion. Its coordinates, relativistic velocity and acceleration can be easily computed:

$$\rho' = e^{\gamma_0} \sqrt{2C_1 \mu_-^0} = \text{const} = \rho_0',$$

$$\varphi' = \varphi_0' + \frac{\Omega_0 c}{2\tilde{a}_0} \ln \frac{\sqrt{1 + \left(\frac{\tilde{a}_0 t'}{c}\right)^2} + \frac{\tilde{a}_0 t'}{c}}{\sqrt{1 + \left(\frac{\tilde{a}_0 t'}{c}\right)^2} - \frac{\tilde{a}_0 t'}{c}}, \tag{2.66}$$

$$z' = \frac{c^2}{\tilde{a}_0} \left[\sqrt{1 + \left(\frac{\tilde{a}_0 t'}{c}\right)^2} - 1 \right].$$

If $\tilde{a}_0 \to 0$, we obtain from (2.66)

$$z' \approx \frac{\tilde{a}_0 (t')^2}{2}, \quad \varphi' \approx \varphi_0' + \Omega_0 t'. \tag{2.67}$$

Thus, a point that is stationary in the reference frame of the non-inertial system (NIS) moves along the axis in the inertial reference frame (IRF) with a velocity of

$$v' = \frac{dz'}{dt'} = \frac{\tilde{a}_0 t'}{\sqrt{1 + \left(\frac{\tilde{a}_0 t'}{c}\right)^2}}$$

and with an acceleration of

$$\tilde{a}' = \frac{d}{dt'}\left[\frac{v'}{\sqrt{1 - \frac{v'^2}{c^2}}}\right] = \tilde{a}_0$$

and simultaneously rotates around the axis with an angular velocity of

$$\frac{d\varphi'}{dt'} = \frac{\Omega_0}{\sqrt{1 + \left(\frac{\tilde{a}_0 t'}{c}\right)^2}}$$

and with an angular acceleration of

$$\frac{d}{dt'}\left[\frac{\frac{d\varphi'}{dt'}}{\sqrt{1 - \frac{v'^2}{c^2}}}\right] = 0$$

along a helical path.

2.5   The Papapetrou Class of Solutions: One-Stationary Soliton Solution

In 1953, Papapetrou obtained the following new class of equations (2.13), (2.15):

$$f_P = C_1 \cdot \frac{\sinh U_0}{\cosh(\psi + U_0)}, \qquad \Phi_P = C_1 \cdot \frac{\cosh \psi}{\cosh(\psi + U_0)} + C_2,$$

$$\omega_P = \frac{1}{C_1}\frac{\partial \chi}{\partial z} \cdot \frac{1}{\sinh U_0} + C_3,$$

$$\tag{2.68}$$

where C_1, C_2, C_3 and U_0 are arbitrary constants and $\psi(\rho, z) = (1/\rho)(\partial\chi/\partial\rho)$ and $\chi(\rho, z)$ are functions satisfying the linear equations (2.9), (2.11), respectively.

This class of solutions can also be obtained with the help of Equations (2.19). Impose now, on the functions entering into system (2.19), the algebraic constraint

$$A_1 B_2 - A_2 B_1 = 0. \tag{2.69}$$

In this case, the function $M(\rho, z)$ can be introduced in the following way:

$$B_1 = MA_1, \quad B_2 = MA_2. \tag{2.70}$$

And one can obtain for M the equation

$$M(1 + M^2)\Delta M = (1 + 2M^2)(\vec{\nabla}M)^2. \tag{2.71}$$

We can seek solution in the form $M = M(\psi)$, where $\Delta\psi = 0$, and after the integration, we get

$$M = \frac{1}{\sinh(\psi + U_0)}, \quad U_0 = \text{const.} \tag{2.72}$$

If M is known, the functions f and Φ can be found from

$$A_1 = \frac{1}{M(M^2 + 1)}\frac{\partial M}{\partial\rho}, \quad A_2 = \frac{1}{M(M^2 + 1)}\frac{\partial M}{\partial z},$$

$$B_1 = \frac{1}{M^2 + 1}\frac{\partial M}{\partial\rho}, \quad B_2 = \frac{1}{M^2 + 1}\frac{\partial M}{\partial z}. \tag{2.73}$$

We have obtained expressions for the functions f and Φ in the form (2.68). Thus, the Papapetrou class of solutions is characterized by the dependents $f = f(\Phi)$.

We note that to find the respective function γ_P one should integrate the equations

$$\frac{\partial\gamma_p}{\partial\rho} = \frac{\rho}{4}\left[\left(\frac{\partial\psi}{\partial\rho}\right)^2 - \left(\frac{\partial\psi}{\partial z}\right)^2\right], \quad \frac{\partial\gamma_p}{\partial z} = \frac{\rho}{2}\frac{\partial\psi}{\partial\rho}\cdot\frac{\partial\psi}{\partial z}. \tag{2.74}$$

1. Consider some special cases:

1-I. $C_1 = 1$, $C_2 = C_3 = 0$, $U_0 \to \infty$. In this case, we have

$$f = e^{-\psi}, \quad \phi = \omega = 0, \quad \Delta\psi = 0, \tag{2.75}$$

and the metric function γ can be obtained from equations (2.74). This is a class of vacuum static Weyl solutions.

1-II. $C_1 = \frac{1}{\sinh U_0}$, $C_2 = -\frac{1}{\sinh U_0 \cosh U_0}$, $C_3 = 0$, $U_0 \to 0$.
Field equations yield

$$f = \frac{2e^{\psi}}{e^{2\psi} + 1}, \quad \Phi = \frac{e^{2\psi} - 1}{e^{2\psi} + 1}, \quad \omega = \frac{\partial\chi}{\partial z}. \tag{2.76}$$

Solutions of this type have been examined by Papapetrou [3]. Physical studies in the field of the Papapetrou solutions (2.76), where performed in Ref. [53].

1-III. $C_1 = \coth U_0$, $C_2 = -\frac{1}{\sinh U_0}$, $C_3 = 0$.
In this case, we obtain the solution

$$f = \frac{1}{\cosh\psi + \sinh\psi \tanh U_0},$$

$$\Phi = -\frac{\sinh\psi}{\cosh U_0} \cdot \frac{1}{\cosh\psi + \sinh\psi \tanh U_0}, \tag{2.77}$$

$$\omega = \frac{1}{\cosh U_0} \cdot \frac{\partial\chi}{\partial z}.$$

Choosing the harmonic function ψ in the form

$$\psi = \ln\frac{x+1}{x-1},$$

we arrive at the Newman–Unti–Tamburino (NUT) solution [54, 55]:

$$f = \frac{x^2 - 1}{x^2 + 2x\tanh U_0 + 1}, \quad \Phi = -\frac{2}{\cosh U_0} \cdot \frac{x}{x^2 + 2x\tanh U_0 + 1},$$

$$\omega = \frac{2k_0 y}{\cosh U_0}, \quad \gamma = \frac{1}{2}\ln\frac{x^2 - 1}{x^2 - y^2}. \tag{2.78}$$

It should be noted that solution (2.78) can be obtained from the Schwarzschild solution $\xi_0 = x$ using the constant phase transformation (2.31):

$$\xi = x\left(\tanh U_0 - i\frac{1}{\cosh U_0}\right). \tag{2.79}$$

In the coordinates r and θ ($x = (r - m)/k_0, y = \cos\theta, k_0^2 = m^2 + l^2$), the metric (2.78) for the NUT solution has the form

$$ds^2 = \left(1 - 2\frac{mr + l^2}{r^2 + l^2}\right)^{-1} dr^2 + (r^2 + l^2)(d\theta^2 + sin^2\theta d\varphi^2)$$

$$- \left(1 - 2\frac{mr + l^2}{r^2 + l^2}\right)(dt - 2l\cos\theta d\varphi)^2, \qquad (2.80)$$

where one can see that there are no asymptotical Minkowski coordinates for this solution. The geometric properties of the NUT spacetime and its physical interpretation are given in Refs. [56–58].

1-IV. If we choose

$$\chi = \int^r \sqrt{\rho^2 + (z - z_1)^2}dz \quad (z_1 = \text{const}),$$

$$\psi = \frac{1}{\rho}\frac{\partial\chi}{\partial\rho} = \ln\frac{\sqrt{\rho^2 + (z - z_1)^2} + (z - z_1)}{\rho},$$

we can obtain the stationary soliton [69, 70]:

$$f = C_1 \cdot \frac{\rho}{\sqrt{\rho^2 + (z - z_1)^2} \cdot \coth U_0 + (z - z_1)},$$

$$\Phi = \frac{C_1\sqrt{\rho^2 + (z - z_1)^2}}{\sqrt{\rho^2 + (z - z_1)^2} \cdot \coth U_0 + (z - z_1)} \cdot \frac{1}{\sinh U_0} + C_2, \qquad (2.81)$$

$$\omega = \frac{1}{C_1}\frac{\sqrt{\rho^2 + (z - z_1)^2}}{\sinh U_0} + C_3.$$

The name "soliton" is applied in general relativity because of the inverse scattering technique used for their obtaining rather than due to specific properties of these solutions. The solution is general relativity; however, it generally does not resemble the classical solitons [71, 72]. It should be noted that the specific geometric and physical properties of the solution (2.81) are described in Refs. [71–90].

2. To conclude this section, we present two simple methods of obtaining the Papapetrou class of solutions:

2-I. Let us rewrite (2.15) in the form

$$\vec{\nabla}\left[\frac{\vec{\nabla}(f^2 + \Phi^2)}{f^2}\right] = 0, \quad \vec{\nabla}\left[\frac{\vec{\nabla}\Phi}{f^2}\right] = 0. \tag{2.82}$$

Choosing $f = f(\psi)$, $\Phi = \Phi(\psi)$ (ψ is the harmonic function) in the above relations, one obtains the following system:

$$\frac{d}{d\psi}\left[\frac{1}{f^2}\frac{d}{d\psi}(f^2 + \Phi^2)\right] = 0, \quad \frac{d}{d\psi}\left[\frac{1}{f^2}\frac{d\Phi}{d\psi}\right] = 0. \tag{2.83}$$

The functions f and Φ should be found from the first-order ordinary differential equations:

$$\frac{d}{d\psi}(f^2 + \Phi^2) = a_0 f^2, \quad \frac{d\Phi}{d\psi} = b_0 f^2 \tag{2.84}$$

(a_0 and b_0 are arbitrary constants).

Integration of (2.84) then yields the expressions for f and Φ:

$$f = C_1\frac{\sinh U_0}{\cosh(\psi + U_0)}, \quad \Phi = C_1\frac{\cosh\psi}{\cosh(\psi + U_0)} + C_2 \tag{2.85}$$

with a special choice of the integration constants.

2-II. The Papapetrou class of solutions (2.85) can also be found with the help of the transformation (2.27).

Setting

$$\tilde{f} = e^{\psi + U_0}, \quad \tilde{\Phi} = 0, \quad A_0 = B_0 = \sinh U_0,$$

$$C_0 = D_0 = 1, \quad E_0 = \cosh U_0,$$

where $U_0 = $ const and ψ is a function satisfying the equation $\Delta\psi = 0$, we obtain (2.85) if one introduces the constants C_1 and C_2.

2.6   The Tomimatsu–Sato Class

This class of solutions can be obtained if the functions f and Φ are related by the formula

$$\frac{\partial f}{\partial x}\frac{\partial f}{\partial y} + \frac{\partial \Phi}{\partial x}\frac{\partial \Phi}{\partial y} = 0. \tag{2.86}$$

It is then possible to introduce the function $M(x, y)$ such that

$$\beta_1 = -\frac{1}{M}\alpha_2, \quad \beta_2 = M\alpha_1, \tag{2.87}$$

where $\alpha_1, \alpha_2, \beta_1$ and β_2 are defined by equations (2.20).

After rather tedious straightforward calculations, one gets for $M(x, y)$ the following ordinary differential equation:

$$\left[\ln \frac{\delta_0^2(\eta - M^2)^2 + (\eta + 1)^2(\eta M' - M)^2}{(\eta^2 + M^2)(\eta - M^2)}\right]' = -\frac{1}{\eta - M^2}, \tag{2.88}$$

where $\eta \equiv (x^2 - 1)/(1 - y^2)$ and a prime denotes $d/d\eta, \delta_0$ is an integrating constant.

1. Let us examine the case $\delta_0 = 1$.

A trivial solution of equation in this case is $M = M_1 = \text{const} = q_0/p_0$, and one can find the functions f and Φ from the equation using the relations

$$\alpha_1 = \frac{2p_0[2p_0x(p_0xx_0 + q_0yy_0 + 1) - x_0(p_0^2x^2 + q_0^2y^2 - 1)]}{(p_0^2x^2 + q_0^2y^2)^2 + 2(p_0^2x^2 + q_0^2y^2 - 1)(p_0xx_0 + q_0yy_0) - 1},$$

$$\alpha_2 = \frac{2q_0[2q_0y(p_0xx_0 + q_0yy_0 + 1) - y_0(p_0^2x^2 + q_0^2y^2 - 1)]}{(p_0^2x^2 + q_0^2y^2)^2 + 2(p_0^2x^2 + q_0^2y^2 - 1)(p_0xx_0 + q_0yy_0) - 1},$$

$$\beta_1 = -\frac{2p_0[2q_0y(p_0xx_0 + q_0yy_0 + 1) - y_0(p_0^2x^2 + q_0^2y^2 - 1)]}{(p_0^2x^2 + q_0^2y^2)^2 + 2(p_0^2x^2 + q_0^2y^2 - 1)(p_0xx_0 + q_0yy_0) - 1},$$

$$\beta_2 = \frac{2q_0[2p_0x(p_0xx_0 + q_0yy_0 + 1) - x_0(p_0^2x^2 + q_0^2y^2 - 1)]}{(p_0^2x^2 + q_0^2y^2)^2 + 2(p_0^2x^2 + q_0^2y^2 - 1)(p_0xx_0 + q_0yy_0) - 1}. \tag{2.89}$$

Thus, we obtain the solution

$$f = \frac{p_0^2 x^2 + q_0^2 y^2 - 1}{(p_0 x + x_0)^2 + (q_0 y + y_0)^2},$$

$$\Phi = -2\frac{p_0 x y_0 - q_0 y x_0}{(p_0 x + x_0)^2 + (q_0 y + y_0)^2},$$

$$(2.90)$$

where p_0, q_0, x_0 and y_0 are real constants satisfying the conditions

$$p_0^2 + q_0^2 = 1, \quad x_0^2 + y_0^2 = 1.$$

According to expressions (2.14) and (2.7), the metric functions ω and γ can be calculated in the form

$$\omega = -2k_0 q_0 \frac{(1 - y^2)(q_0 y y_0 - p_0 x x_0 - 1)}{p_0 (p_0^2 x^2 + q_0^2 y^2 - 1)},$$

$$\gamma = \frac{1}{2}\ln\left(\frac{p_0^2 x^2 + q_0^2 y^2 - 1}{p_0^2 (x^2 - y^2)}\right).$$

$$(2.91)$$

The function ξ calculated using $((2.22), (2.24))$ in this case can be written in the form

$$\xi = (p_0 x + i q_0 y)e^{i\alpha},$$

$$(2.92)$$

where $\cos\alpha = x_0$ and $\sin\alpha = y_0$.

1-I. If we put in (2.90) and (2.91)

$$p_0 = \frac{1 + a_0 b_0}{\sqrt{(1 + a_0^2)(1 + b_0^2)}}, \quad q_0 = \frac{b_0 - a_0}{\sqrt{(1 + a_0^2)(1 + b_0^2)}},$$

$$x_0 = \frac{1 - a_0 b_0}{\sqrt{(1 + a_0^2)(1 + b_0^2)}}, \quad y_0 = \frac{b_0 + a_0}{\sqrt{(1 + a_0^2)(1 + b_0^2)}},$$

$$(2.93)$$

where a_0 and b_0 are arbitrary constants, we can obtain

$$f = \frac{(x^2 - 1)(1 + a_0 b_0)^2 + (y^2 - 1)(b_0 - a_0)^2}{[(1 + a_0 b_0)x + (1 - a_0 b_0)]^2 + [(b_0 - a_0)y + (b_0 + a_0)]^2},$$

$$\Phi = -2\frac{(b_0 + a_0)(1 + a_0 b_0)x - (b_0 - a_0)(1 - a_0 b_0)y}{[(1 + a_0 b_0)x + (1 - a_0 b_0)]^2 + [(b_0 - a_0)y + (b_0 + a_0)]^2}.$$

$$(2.94)$$

The function ξ in this case has the form

$$\xi = \frac{1 + a_0 b_0}{(1 - ia_0)(1 - ib_0)} x + i\frac{b_0 - a_0}{(1 - ia_0)(1 - ib_0)} y. \qquad (2.95)$$

Let as consider some particular case of this solution.

1-II. If we put

$$p_0 = k_0(m^2 + l^2)^{-1/2}, \quad q_0 = a(m^2 + l^2)^{-1/2}, \quad x_0 = m(m^2 + l^2)^{-1/2},$$

$$y_0 = -l(m^2 + l^2)^{-1/2}, \quad k_0^2 = m^2 + l^2 - a^2,$$

where k_0, m, l and a a are arbitrary constants, we can obtain in the coordinates (r, θ) $(x = (r - m)/k_0, y = \cos \theta)$

$$ds^2 = \frac{r^2 + (a\cos\theta - l)^2}{r^2 - 2mr + a^2 - l^2} dr^2 + [r^2 + (a\cos\theta - l)^2]$$

$$\times \left[d\theta^2 + \frac{(r^2 - 2mr + a^2 - l^2)\sin^2\theta d\varphi^2}{r^2 - 2mr + a^2\cos^2\theta - l^2} \right]$$

$$- \left[1 - 2\frac{mr + l(l - a\cos\theta)}{r^2 + (a\cos\theta - l)^2} \right]$$

$$\times \left\{ dt - \left(\frac{2a\sin^2\theta[mr + l(l - a\cos\theta)]}{r^2 - 2mr + a^2\cos^2\theta - l^2} - 2l\cos\theta \right) d\varphi \right\}^2. $$

$$(2.96)$$

This is the Kerr–NUT solution. Its geometric properties Kerr–NUT spacetime are studied in Refs. [59, 91, 92].

1-III. If we put $y_0 = 0$ $(l = 0)$, we obtain the Kerr solution

$$f = \frac{p_0^2 x^2 + q_0^2 y^2 - 1}{(p_0 x + 1)^2 + q_0^2 y^2}, \quad \Phi = \frac{2q_0 y}{(p_0 x + 1)^2 + q_0 y^2},$$

$$\omega = -2k_0 q_0 \frac{(1 - y^2)(p_0 x + 1)}{p_0(p_0^2 x^2 + q_0^2 y^2 - 1)}, \quad \gamma = \frac{1}{2}\ln\left(\frac{p_0^2 x^2 + q_0^2 y^2 - 1}{p_0^2(x^2 - y^2)} \right). $$

$$(2.97)$$

If $k_0^2 = m^2 - a^2$, $p_0 = k_0/m$, $q_0 = a/m$ in the coordinates (r, θ) from (2.96), the Kerr metric takes the form [93].

The potential ξ for the Kerr metric is

$$\xi = p_0 x + i q_0 y, \quad p_0^2 + q_0^2 = 1. \tag{2.98}$$

Robinson [94] showed the existence of a unique family which is asymptotically flat and stationary and has a regular event horizon. The significance of this result is the fact that a rotating black hole is described by the Kerr solution. The Tomimatsu–Sato solutions do not violate this theorem because their coordinate singularities are not enclosed beyond an event horizon. We do not state here the precise conditions of these theorems, which are somewhat technical, and refer the readers to the papers covering the details of the problem.

The geometric properties of the Kerr spacetime and its physical meaning are described in Refs. [95–121]. A more consistent reference list concerning the analysis of the Kerr metric is given in the monograph [122].

1-IV. $q_0 = 0$. In this case, we obtain the NUT solution (2.78).

1-V. $p_0 = x_0, q_0 = -y_0$. Then, we have the solution

$$f = \frac{(x^2 - 1) + a_0^2(y^2 - 1)}{(x + 1)^2 + a_0^2(y - 1)^2}, \quad \Phi = \frac{2a_0(x + y)}{(x + 1)^2 + a_0^2(y - 1)^2}, \tag{2.99}$$

which is known in the world literature as the "extremal" Kerr black hole.

1-VI. $q_0 = y_0 = 0$. In this case, we have

$$f = \frac{x - 1}{x + 1}, \quad \gamma = \frac{1}{2} \ln \frac{x^2 - 1}{x^2 - y^2}, \quad \Phi = \omega = 0, \tag{2.100}$$

i.e., the Schwarzschild solution.

2. If $\delta_0 = 2, 3, \ldots$, one should seek the function M in the form

$$M_i = \frac{Q_i(\eta, p_0, q_0)}{P_i(\eta, p_0, q_0)}, \tag{2.101}$$

where Q_i and P_i are some polynomials.

It should be noted that the function M is related to the Cosgrove functions σ_1 and σ_2 [35, 36] by $M = \sigma_1/\sigma_2$. In Ref. [35], these functions are tabulated for different integer.

Independent of Cosgrove, this case was investigated by Dale [123].

Our functions f and Φ, corresponding to different values of the function M, give us the class of Tomimatsu–Sato (TS) [6, 7] solutions with the distortion parameter δ_0. Yamazaki [9, 10] tried to describe this class of solutions for the case of integer δ_0 in the form of integer series.

We would like to note that the result of building f and Φ if the function M is known makes it possible to obtain solutions slightly different from those of the Tomimatsu–Sato class. Possible generalizations of the TS class of solutions can be obtained using different generating techniques. A physical investigation of the TS class of solutions may be found in Refs. [124–130].

2.7 Variable Separation Technique

In this section, we consider a non-trivial variable separation procedure in the stationary Einstein equations by which some already known solutions are found and also a new one [17].

Let us write down equation (2.23) and its complex conjugate:

$$(\varepsilon + \varepsilon^*)\Delta\varepsilon = 2(\overrightarrow{\nabla}\varepsilon)^2,$$
$$(\varepsilon + \varepsilon^*)\Delta\varepsilon^* = 2(\overrightarrow{\nabla}\varepsilon^*)^2. \tag{2.102}$$

After the substitution $\varepsilon = \varepsilon(\zeta)$, equations (2.102) assume the form

$$[\varepsilon(\zeta) + \varepsilon^*(\zeta^*)]\frac{d\varepsilon}{d\zeta}\Delta\zeta = \left\{2\left(\frac{d\varepsilon}{d\zeta}\right)^2 - [\varepsilon(\zeta) + \varepsilon^*(\zeta^*)]\frac{d^2\varepsilon}{d\zeta^2}\right\}(\overrightarrow{\nabla}\zeta)^2,$$

$$[\varepsilon(\zeta) + \varepsilon^*(\zeta^*)]\frac{d\varepsilon^*}{d\zeta^*}\Delta\zeta^* = \left\{2\left(\frac{d\varepsilon^*}{d\zeta^*}\right)^2 - [\varepsilon(\zeta) + \varepsilon^*(\zeta^*)]\frac{d^2\varepsilon^*}{d\zeta^{*2}}\right\}(\overrightarrow{\nabla}\zeta^*)^2 \tag{2.103}$$

since

$$\overrightarrow{\nabla}\varepsilon = \frac{d\varepsilon}{d\zeta}\overrightarrow{\nabla}\zeta, \quad \Delta\varepsilon = \frac{d\varepsilon}{d\zeta}\Delta\zeta + \frac{d^2\varepsilon}{d\zeta^2}(\overrightarrow{\nabla}\zeta)^2.$$

Impose now an additional condition on equations (2.103):

$$(\vec{\nabla}\zeta)^2 = (\vec{\nabla}\zeta^*)^2. \tag{2.104}$$

Then, by choosing $\varepsilon(\zeta)$, this condition allows us to obtain particular classes of solutions to equation (2.23). Consider some opportunities.

1. $\varepsilon = \zeta$

In this case,

$$\Delta(\zeta - \zeta^*) = 0. \tag{2.105}$$

Its simplest solution $\zeta - \zeta^* = 2ia_0 z$, a_0 being a real constant. Then, we can put

$$\zeta = f(\rho) + ia_0 z, \tag{2.106}$$

where a real function $f(\rho)$ satisfies the equation

$$f\left(\frac{d^2 f}{d\rho^2} + \rho^{-1}\frac{df}{d\rho}\right) = \left(\frac{df}{d\rho}\right)^2 - a_0^2. \tag{2.107}$$

For potentials f and Φ, we finally get

$$f = \begin{cases} \rho C_1^{-1}\sinh[C_1 a_0(C_2 - \ln\rho)], \\ \rho C_1^{-1}\sin[C_1 a_0(\ln\rho - C_2)], \quad \Phi = a_0 z, \\ \rho a_0(\ln\rho - C_2), \end{cases} \tag{2.108}$$

where C_1 and C_2 are real constants.

In the special case $C_1 = a_0^{-1}$, $C_2 = \ln(2a_0^{-1})$, the upper expression for f gives a solution of the form

$$f = 1 - \frac{1}{4}a_0^2\rho^2, \quad \Phi = a_0 z, \tag{2.109}$$

which is a stationary vacuum analogue of Bonnor's solution for a uniform electric or magnetic universe [131].

2. $\varepsilon = (\zeta - 1)(\zeta + 1)^{-1}$, $\zeta = \xi$.

In this case, field equations take the form (2.25)

$$(\xi\xi^* - 1)\Delta\xi = 2\xi^*(\vec{\nabla}\xi)^2, \quad (\xi\xi^* - 1)\Delta\xi^* = 2\xi(\vec{\nabla}\xi^*)^2,$$
$$\tag{2.110}$$

where from we get

$$\Delta(\xi^2 - \xi^{*2}) = 0. \tag{2.111}$$

2-I. Again, choosing the simplest solution of this equation $\xi^2 - \xi^{*2} = 2ia_0 z$ and putting $a_0 = 2pq$ (the constants p and q satisfy $p^2 + q^2 = 1$), we separate the variables by passing to the prolate ellipsoidal coordinates (x, y):

$$\xi = px + iqy, \quad \xi^* = px - iqy. \tag{2.112}$$

One finally obtains

$$f = \frac{p^2 x^2 + q^2 y^2 - 1}{(px + 1)^2 + q^2 y^2}, \quad \Phi = \frac{2qy}{(px + 1)^2 + q^2 y^2}.$$

This relation determines the already considered Kerr solution.

2-II. Consider the following solution of the equation:

$$\xi = \frac{1}{2}(p + iq)(x + y)K_+ + \frac{1}{2}(p - iq)(x - y)K_-, \tag{2.113}$$

where

$$K_\pm = \sqrt{1 + \frac{z_0'}{(x \pm y)^2}(z_0' + 2xy \pm 2)},$$

$$z_0' = \frac{z_0}{k_0}, \quad k_0 = pm,$$

$$p = \sqrt{1 - \frac{a^2}{m^2}}, \quad q = \frac{a}{m}, \quad p^2 + q^2 = 1.$$

We arrive at the solution that corresponds to a shift of the Kerr solution along the z axis by z_0 in the canonical Weyl coordinates (ρ, z). For $z_0 = 0$, we obtain (2.112).

For $a = 0$, we arrive at the Schwarzschild solution with a shift $z \to z + z_0$, which in curvature coordinates takes the form

$$f = \frac{(r - m)(K_+ + K_-) + m\cos\theta \cdot (K_+ - K_-) - 2m}{(r - m)(K_+ + K_-) + m\cos\theta \cdot (K_+ - K_-) + 2m}. \tag{2.114}$$

For $z_0 = 0$, the solution (2.114) reduces to the Schwarzschild solution.

3. $\varepsilon = e^{\zeta}$.

For this substitution, we find that

$$\Delta(\zeta + \zeta^*) = 0 \tag{2.115}$$

and a solution in the form $\zeta + \zeta^* = 2a_0 z$ can be found by separating the variables

$$\zeta = a_0 z + iM(\rho), \quad \zeta^* = a_0 z - iM(\rho), \tag{2.116}$$

where the real function $M(\rho)$ satisfies the equation

$$\frac{d^2 M}{d\rho^2} + \rho^{-1}\frac{dM}{d\rho} = \tan\left[a_0^2 - \left(\frac{dM}{d\rho}\right)^2\right]. \tag{2.117}$$

The expressions for f and Φ have the form

$$f = e^{a_0 z}\cos M, \quad \Phi = e^{a_0 z}\sin M. \tag{2.118}$$

Solutions of this type were examined by Marek [26] and were not obtained in an explicit form.

4. $\zeta = \tanh(\zeta/2)$.

With this substitution equation, it follows the equation

$$\Delta(\zeta^* - \zeta) = 0 \tag{2.119}$$

enabling us to separate the variables in the following way:

$$\zeta = F(\rho) + ia_0 z, \quad \zeta^* = F(\rho) - ia_0 z. \tag{2.120}$$

The real function $F(\rho)$ satisfies the ordinary differential equation

$$\frac{d^2 F}{d\rho^2} + \rho^{-1}\frac{dF}{d\rho} = \cot\left[\left(\frac{dF}{d\rho}\right)^2 - a_0^2\right], \tag{2.121}$$

and the expressions for f and Φ have the form

$$f = \frac{\sinh F}{\cosh F + \cos a_0 z}, \quad \Phi = \frac{\sin a_0 z}{\cosh F + \cos a_0 z}. \tag{2.122}$$

The substitutions

$$F = \ln\coth\frac{\lambda}{4}, \quad \rho^2 = 4a_0^{-2}e^{u} \tag{2.123}$$

transform (2.121) to the equation

$$\frac{d^2\lambda}{du^2} - e^u \sinh\lambda = 0, \qquad (2.124)$$

which can be investigated with the aid of infinite series [132].

For $a_0 = 0$, the metric determined by the relations (2.121), (2.122) becomes flat.

References

[1] T. Lewis, *Proc. R. Soc. Lond.* A 136, 176 (1932).
[2] W.J. Van Stockum, *Proc. R. Soc. Edinb.* A 57, 135 (1937).
[3] A. Papapetrou, *Ann. Phys.* B 12, 309 (1953).
[4] F.J. Ernst, *Phys. Rev.* 167(5), 1175 (1968).
[5] R.P. Kerr, *Phys. Rev. Lett.* 11(5), 1776 (1963).
[6] A. Tomimatsu, H. Sato, *Phys. Rev. Lett.* 29(19), 1344 (1972).
[7] A. Tomimatsu, H. Sato, *Prog. Theor. Phys.* 50(1), 95 (1973).
[8] W. Kinnersley, in G. Shaviv, J. Rosen (eds.), *Proceedings of 7th International Conference on General Relativity and Gravitation*, Vol. 109, Wiley, New York (1975).
[9] M. Yamazaki, *Prog. Theor. Phys.* 57, 1951 (1977).
[10] M. Yamazaki, *J. Math. Phys.* 18, 2502 (1977).
[11] S. Hori, *Prog. Theor. Phys.* 95(6), 1097 (1996).
[12] C.M. Cosgrove, *J. Phys. A: Math. Gen.* A10, 1481 (1977).
[13] C.M. Cosgrove, *J. Phys. A: Math. Gen.* A10, 2093 (1977).
[14] F.J. Ernst, *J. Math. Phys.* 18, 233 (1977).
[15] C. Hoenselaers, *J. Phys. A: Math Gen.* A11(4), 75 (1978).
[16] P.V. Pryse, *Class. Quantum Grav.* 10, 163 (1993).
[17] T.I. Gutsunaev, V.S. Manko, *Jzv. Vuzov Fiz.* 4, 116 (1985).
[18] S. Persides, B.C. Xanthopoulos, *J. Math. Phys.* 29, 674 (1988).
[19] R.B. Hoffman, *J. Math, Phys.* 10, 953 (1969).
[20] D. Cox, W. Kinnersley, *J. Math. Phys.* 20, 1225 (1979).
[21] M.G. Tseitlin, *Teor. Matem. Fiz.* 64, 51 (1985).
[22] J.N. Islam, *Math. Proc. Camb. Phil. Soc.* 79, 161 (1976).
[23] J.N. Islam, *Gen. Rel. Grav.* 7, 809 (1976).
[24] F.J. Ernst, *J. Math. Phys.* 15, 1409 (1974).
[25] H. Levy, *Nuovo Cim.* 56B, 253 (1968).
[26] J.J.J. Marek, *Proc. Camb. Phil. Soc.* 64, 167 (1968).
[27] R. Geroch, *J. Math. Phys.* 12, 918 (1968).
[28] D. Kramer, G. Neugebauer, *Commun. Math. Phys.* 10, 132 (1968).
[29] R. Catenacci, D. Alonso, *J. Math. Phys.* 17, 2232 (1976).

[30] Y. Tanabe, *J. Math. Phys.* 20, 1486 (1979).

[31] C.M. Cosgrove, *J. Math. Phys.* 21, 2417 (1980).

[32] C.M. Cosgrove, *J. Math. Phys.* 22(11), 2624 (1981).

[33] C.M. Cosgrove, *J. Math Phys.* 23, 615 (1982).

[34] J. Ehlers, Konstructionen und Charakterisierung der Einsteinschen Gravitationsfeldgleichungen, Dissertation, Hamburg (1957).

[35] C.M. Cosgrove, *J. Phys. A: Math. Gen.* 11, 2389 (1978).

[36] C.M. Cosgrove, *J. Phys. A: Math. Gen.* 11, 2405 (1978).

[37] R. Geroch, *J. Math. Phys.* 11, 1955 (1970).

[38] R. Geroch, *J. Math. Phys.* 11, 2580 (1970).

[39] O. Hansen, *J. Math. Phys.* 15, 46 (1974).

[40] C. Hoeselaers, in H. Sato, T. Nakamura (eds.), *Gravitational Collapse and Relativity. Proceedings of 14th Yamada Conference*, Kyoto, Japan, World Scientific, Singapore (1986), p. 176.

[41] K.S. Thome, *Rev. Mod. Phys.* 52, 299 (1980).

[42] R. Beig, W. Simon, *Proc. R. Soc. Lond.* A 376, 333 (1981).

[43] Y. Gursel, *Gen. Rel. Grav.* 12, 1003 (1983).

[44] R. Beig, W. Simon, *J. Math. Phys.* 24, 1163 (1983).

[45] D. Kramer, H. Stephani, M.A.H. Mac Callum, E. Herit, *Exact Solutions of Einstein's Field Equations*, Cambridge University Press, Cambridge (1980).

[46] E. Kasner, *Am. J. Math.* 43, 217 (1921).

[47] V. Narlikar, K. Karmarkar, *Curr. Sci.* 15, 69 (1946).

[48] A. Kellner, 1-dimensionale Gravitationsfelder, Dissertation, Gottingen (1975).

[49] A. Harvey, *Gen. Rel. Grav.* 21(10), 1021 (1989).

[50] K.D. Krori, R. Bhattacharjee, *J. Math. Phys.* 31, 147 (1990).

[51] M.A.H. Mac Callum, *Wiss. Z. Friedrich-Schiller Univ. Jena, Naturwiss. R.* 39, 102 (1990).

[52] R.B. Hoffman, *Phys. Rev.* 182, 1361 (1969).

[53] D. Phan, *Gen. Rel. Grav.* 23, 269 (1991).

[54] E.T. Newman, L. Tamburino, T.W.J. Unti, *J. Math. Phys.* 4, 915 (1963).

[55] E.T. Newman, T.W.J. Unti, *J. Math. Phys.* 4, 1467 (1963).

[56] A.H. Taub, *Ann. Math.* 53, 475 (1951).

[57] C.W. Misner, *J. Math. Phys.* 4, 924 (1963).

[58] W.B. Bonnor, N.S. Swaminarayan, *Z. Phys.* 77, 240 (1965).

[59] M. Demianski, E.T. Newman, *Bull. Akad. Polon. Sci. Ser. Math. Astron. Phys.* 14, 653 (1966).

[60] C.W. Misner, in J. Ehlers (ed.), *Lectures in Applied Mathematics*, AMS, Providence (1967).

[61] C.W. Misner, A.H. Taub, *Zh. Eksper. Tear. Fiz.* 55, 233 (1968).

[62] W.B. Bonnor, *Proc. Camb. Phil. Soc.* 66, 145 (1969).

[63] A. Sackfield, *Proc. Camb. Phil. Soc.* 70, 89 (1971).

[64] R. Gautreau, R.B. Hoffman, *Phys. Lett.* A39, 75 (1972).

[65] C. Reina, A. Treves, *J. Math. Phys.* 16(4), 834 (1975).

[66] M.P. Ryan, L.C. Shepley, *Homogeneous Relativistic Cosmologies*, Princeton University Press, Princeton (1975).

[67] S.T.C. Siklos, *Phys. Lett.* A59, 173 (1976).

[68] R.L. Zimmerman, B.Y. Shahir, *Gen. Rel. Grav.* 21, 821 (1989).

[69] V.A. Belinsky, V.E. Zakharov, *Sov. Phys. JETP* 48, 985 (1978).

[70] V.A. Belinsky, V.E. Zakharov, *Sov. Phys. JETP* 50, 1 (1979).

[71] J. Ibanez, E. Verdaguer, *Phys. Rev.* D 31, 251 (1985).

[72] P.T. Boyd, J.M. Centrella, S.A. Klasky, *Phys. Rev.* D 43, 79 (1991).

[73] R.T. Jantzen, *Nuovo Cim.* B 59, 287 (1980).

[74] G.A. Alekseev, *Sov. Phys. Dokl.* 26, 158 (1981).

[75] G. Neugebauer, *Phys. Lett.* A86, 91 (1981).

[76] V. Belinsky, M. Francaviglia, *Gen. Rel. Grav.* 14, 213 (1982).

[77] E. Verdaguer, *Phys. A: Math. Gen.* 15, 1261 (1982).

[78] M. Yamazaki, *Phys. Lett.* 50, 1027 (1983).

[79] B.J. Carr, E. Verdaguer, *Phys. Rev.* D28, 2995 (1983).

[80] J. Ibanez, E. Verdaguer, *Phys. Rev. Lett.* 51, 1313 (1983).

[81] A.P. Veselov, *Theor. Math. Phys.* 54, 155 (1983).

[82] R.J. Gleiser, *Gen. Relat. Grav.* 16(11), 1077 (1984).

[83] P.S. Letelier, *J. Math. Phys.* 26(3), 467 (1985).

[84] K.C. Das, *Phys. Rev.* D31(4), 927 (1985).

[85] G.A. Alekseev, *Gravit. Electromagn.* 4, 3 (1989).

[86] A. Dagotto, R. Gleiser, G. Gonzalez, J. Pullin, *Phys. Lett.* A146, 15 (1990).

[87] S.I. Tertychniy, *Class. Quant. Grav.* 7(8), 1345 (1990).

[88] V. Belinsky, *Phys. Rev.* D 44, 3109 (1991).

[89] Y. Wang, Z. He, *J. Shanghai Jiaotong Univ.* 28, 62 (1994).

[90] A. Nakamura, *J. Phys. Soc. Jpn.* 63(3), 1214 (1994).

[91] G.E. Sneddon, *J. Math. Phys.* 16, 740 (1975).

[92] Y. Wang, *J. Beijing Normal Univ. Nat. Sci.* 27(3), 280 (1991).

[93] R.H. Boyer, R.W. Lindquist, *J. Math. Phys.* 8, 265 (1967).

[94] D.C. Robinson, *Phys. Rev. Lett.* 34, 905 (1975).

[95] R.H. Boyer, T.G. Price, *Proc. Camb. Phil. Soc.* 61, 531 (1965).

[96] W. Israel, *Phys. Rev.* 164, 1776 (1967).

[97] W.C. Hernandez, *Phys. Rev.* 159, 1070 (1967).

[98] B. Carter, *Phys. Rev.* 174, 1559 (1968).

[99] G. Debney, R.P. Kerr, A. Shild, *J. Math. Phys.* 10, 1842 (1969),

[100] W. Kinnersley, *J. Math. Phys.* 10, 1195 (1969).

[101] W. Israel, *Phys. Rev.* D2, 641 (1970).

[102] B. Carter, *Phys. Rev. Lett.* 26, 331 (1971).

[103] B. Carter, in *Black Holes*, Les Hourches Lectures, Gordon and Breach, New York (1972).

[104] E. Herlt, *Wiss. Z. Univ. Jena Math.-Nat. R.* 21, 19 (1972).

[105] S.W. Hawking, *Commun. Math. Phys.* 25, 152 (1972).

[106] S.W. Hawking, G.F.R. Ellis, *The Large Scale Structure of Spacetime*, Cambridge University Press, Cambridge (1973).

[107] J.M. Stewart, M. Nalker, *Black Holes: The Outside Story*, Springer Tracts in Modern Physics, Vol. 69, Springer-Verlag, Berlin (1973).

[108] D. Cox, E. Flaherty, *Commun. Math. Phys.* 47, 75 (1976).

[109] S. Chandrasekhar, *Proc. R. Soc. Lond.* A 358, 405 (1978).

[110] E. Herlt, *Gen. Relat. Grav.* 9, 711 (1978).

[111] E. Herlt, *Gen. Relat. Grav.* 11, 337 (1979).

[112] S.K. Bose, *An Introduction to General Relativity*, Wiley Eastem, New Dehli (1980).

[113] G. Bergqvist, M. Ludvigsen, *Gen. Relat. Grav.* 21, 1171 (1989).

[114] R. Quintana, *Appl. Math. Lett.* 3(1), 51 (1990).

[115] G. Bergqvist, M. Ludvigsen, *Class. Quant. Grav.* 8(4), 697 (1991).

[116] I.G. Fichtengolts, *Dokl. Akad. Nauk SSSR* 317, 355 (1991).

[117] A. Nakamura, Y. Ohta, *J. Phys. Soc. Jpn.* 60, 1835 (1991).

[118] W. Rindler, V. Perlic, *Gen. Relat. Grav.* 22, 1067 (1990).

[119] H. Bondy, W. Rindler, *Gen. Relat. Grav.* 23, 487 (1991).

[120] S. Nitta, A. Takahashi, A. Tomimatsu, *Phys. Rev.* D 44, 2295 (1991).

[121] K. Hayaski, T. Samura, *Phys. Rev.* D 50, 3666 (1994).

[122] S. Chandrasekhar, *The Mathematical Theory of Black Holes*, Oxford University Press, Oxford (1983).

[123] P. Dale, *Proc. R. Soc. Lond.* A 362, 463 (1978).

[124] G.W. Gibbon, R.A. Russel-Clark, *Phys. Rev. Lett.* 30, 398 (1973).

[125] W. Kinnersley, E.F. Kelley, *J. Math. Phys.* 15, 2121 (1974).

[126] Y. Tanabe, *Prog. Theor. Phys.* 52, 727 (1974).

[127] F.J. Ernst, *J. Math. Phys.* 17, 1091 (1976).

[128] J.E. Economou, *J. Math. Phys.* 17, 1095 (1976).

[129] C. Hoenselaers, F.J. Ernst, *J. Math. Phys.* 24, 1817 (1983).

[130] Z. Perjes, *J. Math. Phys.* 30, 2197 (1989).

[131] W.B. Bonnor, *Proc. Phys. Soc. Lond.* A 66, 145 (1953).

[132] M. Chini, *Giornale Math.* 58, 35 (1920).

[133] A.A. Shaideman, Generation of vacuum axially symmetric solutions to the equations of General Relativity using stationary Euclidons, PhD thesis, Moscow (2005) (in Russian).

Chapter 3

New Methods of Generating Stationary Vacuum Einstein Fields

3.1 Introduction

A study of methods of generating exact solutions, based on internal symmetries of the Einstein equations, has been attracting the attention of gravitation physicists.

The publications by Ehlers [1], Ozvath [2] and Harrison [3] pioneer these investigations. Subsequently, one can distinguish three approaches to this problem.

The group-theoretic technique, which is very useful in construction of the new metrics containing an arbitrary large number of parameters, was introduced by Geroch [4], Kinnersley [5] and Maison [6] and used in Refs. [7–20]. The main achievements of this approach are related with the group of continuous transformations of the Ernst equation, with the Hoenselaers–Kinnersley–Xanthopoulos (HKX) transformations [13, 14], with the aid of which one can obtain a set of asymptotically flat stationary vacuum metrics [21–43].

The second approach, introduced originally by Belinsky and Zakharov [44, 45], is based on the application of the inverse scattering method to Einstein's equations. At this point, one can obtain new results in general relativity [46–65].

Finally, the last approach to generate new solutions from old was founded on the applications of the Backlund transformations. For the first time in general relativity, the Backlund transformations were introduced by Harrison [66] and Neugebauer [67], and their possible usage can be found in Refs. [68–76].

Although these approaches are independent, they all are mathematically equivalent, which was shown by Cosgrove [77–79].

The major result of the applications of techniques discussed above is the construction of the nonlinear superposition of N-Kerr solutions aligned along their common rotational axis. Therewith, the Tomimatsu–Sato metric has been recognized to be a special case of this solution [80–100].

Furthermore, the stationary vacuum problem can be reduced to the one singular equation [101]. The application of the technique of the generalized Riemannian problem to general relativity yields a large class of new solutions [102].

In Ref. [103], the method of the variation of constants was proposed. With the help of this technique, one can be brought to the nonlinear superposition of the Kerr spacetime, with an arbitrary vacuum field [104–107].

In Ref. [108], the nonlinear superposition of the stationary Euclidon solution with an arbitrary axially symmetric stationary gravitational field on the basis of the method of variation of parameters was constructed. In Ref. [109], stationary soliton solution of the Einstein equations was generalized to the case of a stationary seed metric. The formulae obtained in Refs. [108, 109] have a simple and compact form, permitting an effective nonlinear "addition" of the solutions. The Euclidon method propounded in Ref. [108] is used in Ref. [110].

In this chapter, the both methods [108, 109] are considered again and some new applications are presented.

3.2   The Euclidon Method

Let $f^0(\rho, z)$, $\Phi^0(\rho, z)$, $\omega^0(\rho, z)$ be some arbitrary solution of the equations

$$f\Delta f = (\vec{\nabla} f)^2 - (\vec{\nabla}\Phi)^2, \quad f\Delta\Phi = 2\vec{\nabla}\Phi\vec{\nabla} f,$$

$$\frac{\partial\omega}{\partial\rho} = \frac{\rho}{f^2}\frac{\partial\Phi}{\partial z}, \quad \frac{\partial\omega}{\partial z} = -\frac{\rho}{f^2}\frac{\partial\Phi}{\partial\rho}. \tag{3.1}$$

We call (f^0, Φ^0, ω^0) the seed solution.

One of the simplest solutions of equation (3.1) is the one-stationary Euclidon solution:

$$f_E = \frac{(z - z_1) + \sqrt{\rho^2 + (z - z_1)^2}\,\tanh U_0}{C_1},$$

$$\Phi_E = \frac{\sqrt{\rho^2 + (z - z_1)^2}}{C_1 \cosh U_0} + C_2, \tag{3.2}$$

$$\omega_E = C_1 \frac{\sqrt{\rho^2 + (z - z_1)^2}}{(z - z_1) + \sqrt{\rho^2 + (z - z_1)^2}\,\tanh U_0} \cdot \frac{1}{\cosh U_0} + C_3,$$

where U_0, C_1, C_2, and C_3 are arbitrary constants.

Straightforward calculations show that solution (3.2) turns all components of the Riemann–Christoffel curvature tensor to zero. Thus, it makes sense to call solution (3.2) a Euclidon solution. One can say that it characterizes some relativistic non-inertial frame of reference in flat spacetime. Nevertheless, solution (3.2) allows one to generate solutions describing curved spacetime.

In order to make a nonlinear "composition" of the solution (3.2) with the seed solution, we use the method of variation of parameters. We consider C_1, C_2, C_3 and U_0 in the solution (3.2) to be functions

$$C_1 \to f^0(\rho, z), \quad C_2 \to \omega^0(\rho, z), \quad C_3 \to \Phi^0(\rho, z), \quad U_0 \to U(\rho, z). \tag{3.3}$$

In this case, we have

$$f = \frac{(z - z_1) + \sqrt{\rho^2 + (z - z_1)^2}\,\tanh U(\rho, z)}{f^0},$$

$$\Phi = \frac{\sqrt{\rho^2 + (z - z_1)^2}}{f^0 \cdot \cosh U(\rho, z)} + \omega^0(\rho, z),$$

$$\omega = f^0 \frac{\sqrt{\rho^2 + (z - z_1)^2}}{(z - z_1) + \sqrt{\rho^2 + (z - z_1)^2}\,\tanh U(\rho, z)} \cdot \frac{1}{\cosh U(\rho, z)}$$

$$+ \Phi^0(\rho, z). \tag{3.4}$$

A substitution of (3.4) into equations (3.1) leads to the following set of first-order differential equations for the unknown function

$U(\rho, z)$:

$$\sqrt{\rho^2 + (z - z_1)^2}\,\frac{\partial U}{\partial \rho}$$

$$= (z - z_1)\frac{1}{f^0}\frac{\partial f^0}{\partial \rho} + \frac{\rho}{f^0}\frac{\partial f^0}{\partial z} + \Big[(z - z_1)\sinh U$$

$$+ \sqrt{\rho^2 + (z - z_1)^2} \cdot \cosh U\Big]\frac{1}{f^0}\frac{\partial \Phi^0}{\partial \rho} + \rho\frac{\sinh U}{f^0}\frac{\partial \Phi^0}{\partial z}, \quad (3.5)$$

$$\sqrt{\rho^2 + (z - z_1)^2}\,\frac{\partial U}{\partial z}$$

$$= -\frac{\rho}{f^0}\frac{\partial f^0}{\partial \rho} + \frac{z - z_1}{f^0}\frac{\partial f^0}{\partial z} + \Big[(z - z_1)\sinh U$$

$$+ \sqrt{\rho^2 + (z - z_1)^2} \cdot \cosh U\Big]\frac{1}{f^0}\frac{\partial \Phi^0}{\partial z} - \rho\frac{\sinh U}{f^0}\frac{\partial \Phi^0}{\partial \rho}.$$

It is readily seen that the integrability condition for equations (3.5) is satisfied.

Equations (3.5) are nonlinear. The problem of linearization of system (3.5) can be solved by the substitution $U(\rho, z) = \ln[a(\rho, z)/b(\rho, z)]$. Then, we have the following set of linear equations for the functions $a(\rho, z)$ and $b(\rho, z)$:

$$2f^0\frac{\partial a}{\partial \rho} = a\left(\hat{L}_1 + \frac{\partial}{\partial \rho}\right)f^0 - b\left(\hat{L}_1 - \frac{\partial}{\partial \rho}\right)\Phi^0,$$

$$2f^0\frac{\partial a}{\partial z} = -a\left(\hat{L}_2 - \frac{\partial}{\partial z}\right)f^0 + b\left(\hat{L}_2 + \frac{\partial}{\partial z}\right)\Phi_0,$$

$$2f^0\frac{\partial b}{\partial \rho} = -b\left(\hat{L}_1 - \frac{\partial}{\partial \rho}\right)f^0 - a\left(\hat{L}_1 + \frac{\partial}{\partial \rho}\right)\Phi^0,$$

$$2f^0\frac{\partial b}{\partial z} = b\left(\hat{L}_2 + \frac{\partial}{\partial z}\right)f^0 + a\left(\hat{L}_2 - \frac{\partial}{\partial z}\right)\Phi^0, \quad (3.6)$$

where $\hat{L}_1$ and $\hat{L}_2$ are the linear operators:

$$\sqrt{\rho^2 + (z - z_1)^2}\,\hat{L}_1 = (z - z_1)\frac{\partial}{\partial \rho} + \rho\frac{\partial}{\partial z},$$

$$\sqrt{\rho^2 + (z - z_1)^2}\,\hat{L}_2 = \rho\frac{\partial}{\partial \rho} - (z - z_1)\frac{\partial}{\partial z}.$$

Thus, the problem of nonlinear superposition of the solution (3.4) with an arbitrary axially symmetric stationary solution of the Einstein equations reduces to integrating a set of first-order linear differential equations. Methods of treating this problem are well studied.

3.3 Application of the Euclidon Method: Two-Stationary Euclidon Solution (The Kerr–NUT Solution)

(a) It is convenient to go over to prolate ellipsoidal coordinates (x, y), which are connected to the Weyl coordinates (ρ, z) by the relations

$$\rho = k_0\sqrt{(x^2 - 1)(1 - y^2)}, \quad z = k_0 xy$$

(k_0 – a real constant).

In terms of the coordinates (x, y), the equations (3.1), (3.4) and (3.5) are rewritten as ($z_1 = k_0$)

$$\frac{\partial \omega^0}{\partial x} = \frac{k_0(1 - y^2)}{(f^0)^2}\frac{\partial \Phi^0}{\partial y}, \quad \frac{\partial \omega^0}{\partial y} = -\frac{k_0(x^2 - 1)}{(f^0)^2}\frac{\partial \Phi^0}{\partial x}, \tag{3.7}$$

$$f = \frac{k_0}{f^0}\left[xy - 1 + (x - y)\tanh U\right], \quad \Phi = \frac{k_0}{f^0}\frac{(x - y)}{\cosh U} + \omega^0, \tag{3.8}$$

$$\omega = f^0\frac{(x - y)}{xy - 1 + (x - y)\tanh U}\frac{1}{\cosh U} + \Phi^0$$

and

$$\frac{\partial U}{\partial x} = \left(\frac{xy - 1}{x - y}\right)\frac{1}{f^0}\frac{\partial f^0}{\partial x} + \left(\frac{1 - y^2}{x - y}\right)\frac{1}{f^0}\frac{\partial f^0}{\partial y}$$

$$+ \left[\cosh U + \left(\frac{xy - 1}{x - y}\right)\sinh U\right]\frac{1}{f^0}\frac{\partial \Phi^0}{\partial x} + \left(\frac{1 - y^2}{x - y}\right)\frac{\sinh U}{f^0}\frac{\partial \Phi^0}{\partial y},$$

$$\frac{\partial U}{\partial y} = -\left(\frac{x^2 - 1}{x - y}\right)\frac{1}{f^0}\frac{\partial f^0}{\partial x} + \left(\frac{xy - 1}{x - y}\right)\frac{1}{f^0}\frac{\partial f^0}{\partial y}$$

$$+ \left[\cosh U + \left(\frac{xy - 1}{x - y}\right)\sinh U\right]\frac{1}{f^0}\frac{\partial \Phi^0}{\partial y} - \left(\frac{x^2 - 1}{x - y}\right)\frac{\sinh U}{f^0}\frac{\partial \Phi^0}{\partial x}.$$

$$\tag{3.9}$$

Choosing the function in the form of

$$U = \ln \frac{a}{\sqrt{f^0}} - \ln \frac{b}{\sqrt{f^0}}, \qquad (3.10)$$

we arrive at the following system of linear differential equations for the functions

$$2f^0 \frac{\partial a}{\partial x} = a(1+y)\hat{L}_1 f^0 + b(1-y)\hat{L}_2 \Phi^0,$$

$$2f^0 \frac{\partial a}{\partial y} = -a(x-1)\hat{L}_2 f^0 + b(x+1)\hat{L}_1 \Phi^0,$$

$$2f^0 \frac{\partial b}{\partial x} = b(1-y)\hat{L}_2 f^0 - a(1+y)\hat{L}_1 \Phi^0,$$

$$2f^0 \frac{\partial b}{\partial y} = b(x+1)\hat{L}_1 f^0 + a(x-1)\hat{L}_2 \Phi^0. \qquad (3.11)$$

The operators take the form

$$(x-y)\hat{L}_1 = (x-1)\frac{\partial}{\partial x} + (1-y)\frac{\partial}{\partial y},$$

$$(x-y)\hat{L}_2 = (x+1)\frac{\partial}{\partial x} - (1+y)\frac{\partial}{\partial y}.$$

1. Let us consider the seed metric $(f^0, \ \Phi^0, \ \omega^0)$ of the form

$$f^0 = (x+1)^\delta (x-1)^{1-\delta}(1+y)^{1-\gamma}(1-y)^\gamma, \quad \Phi^0 = \omega^0 = 0, \quad (3.12)$$

where δ and γ are arbitrary constants. In this case, we have from (3.8)

$$\lim_{U \to \infty} f = \left(\frac{x-1}{x+1}\right)^\delta \left(\frac{1+y}{1-y}\right)^\gamma. \qquad (3.13)$$

As one can easily see, the expression for f given by (3.13) satisfies the static equation

$$\frac{\partial}{\partial x}\left[(x^2-1)\frac{\partial \ln f}{\partial x}\right] + \frac{\partial}{\partial y}\left[(1-y^2)\frac{\partial \ln f}{\partial y}\right] = 0.$$

In the case $\gamma = 0$, the expression (3.13) reduces to the Zipoy solution.

If we choose the seed metric in form (3.12), then the unknown function $U(x, y)$ for this case can be found from the equations

$$\frac{\partial U}{\partial x} = \frac{\delta}{x + 1} + \frac{\delta - 1}{x - 1} + 2\frac{1 - \gamma - \delta}{x - y},$$

$$\frac{\partial U}{\partial y} = \frac{\gamma - 1}{1 + y} - \frac{\gamma}{1 - y} + 2\frac{\delta + \gamma - 1}{x - y}. \tag{3.14}$$

And after the integration in (3.14), we come to the following function:

$$U(x, y) = \ln\left[\frac{(x - y)^{2(1 - \gamma - \delta)}}{\alpha} \cdot \frac{(x^2 - 1)^{\delta}}{x - 1} \cdot \frac{(1 - y^2)^{\gamma}}{1 + y}\right], \tag{3.15}$$

where α is an integration constant. Note that in the limiting case $\alpha \to 0$, the solution

$$f = \frac{xy - 1 + (x - y)\tanh U}{f^0} \tag{3.16}$$

reduced to (3.13).

Substituting (3.12) and (3.15) in (3.8), we arrive at the solution

$$\tilde{f} = \left(\frac{x - 1}{x + 1}\right)^{\delta}\left(\frac{1 + y}{1 - y}\right)^{\gamma} \cdot \frac{\tilde{A}}{\tilde{B}},$$

$$\tilde{\Phi} = \frac{2\alpha\tilde{C}}{\tilde{B}}, \quad \tilde{\omega} = \frac{2k_0'\alpha\tilde{D}}{\tilde{A}}, \tag{3.17}$$

where

$$\tilde{A} = (x - y)^{4(1 - \gamma - \delta)} - \alpha^2(x^2 - 1)^{-2\delta + 1}(1 - y^2)^{-2\gamma + 1},$$

$$\tilde{B} = (x - y)^{4(1 - \gamma - \delta)} + \alpha^2(x + 1)^{-2\delta}(x - 1)^{-2\delta + 2}(1 + y)^{-2\gamma + 2}$$

$$\times (1 - y)^{-2\gamma},$$

$$\tilde{C} = (x - y)^{1 + 2(1 - \gamma - \delta)}(x + 1)^{-2\delta}(1 - y)^{-2\gamma},$$

$$\tilde{D} = (x - y)^{1 + 2(1 - \gamma - \delta)}(x - 1)^{-2\delta + 1}(1 + y)^{-2\gamma + 1}.$$

Using the transformational theorem, we can construct s new class of solutions

$$f = \frac{(ad + bc)\tilde{f}}{(c - d\tilde{\Phi})^2 + d^2\tilde{f}^2},$$

$$\Phi = \frac{-ac + (ad - bc)\tilde{\Phi} + bd(\tilde{f}^2 + \tilde{\Phi}^2)}{(c - d\tilde{\Phi})^2 + d^2\tilde{f}^2},$$

(3.18)

where a, b, c and d are arbitrary constants.

Setting

$$P_0 = \frac{c^2 + d^2}{c^2 - d^2}, \quad Q_0 = \frac{2cd}{c^2 - d^2}, \quad (P_0^2 - Q_0^2 = 1),$$

$$P_0' = \frac{ac - bd}{ac + bd}, \quad Q_0' = \frac{ad - bc}{ac + bd},$$

$$R_0' = \frac{ad + bc}{ac + bd}, \quad P_0'^2 + R_0'^2 = Q_0'^2 + 1,$$

we finally have [111]

$$f = \frac{A}{B}, \quad \Phi = -\frac{C}{B}, \quad \omega = \frac{2k_0}{R_0'} \cdot \frac{K_+ - K_- + L}{M} - \frac{2k_0 Q_0}{R_0'}(\delta y - \gamma x),$$

(3.19)

where

$$A = R_0' \cdot f^0[xy - 1 + (x - y)\tanh U],$$

$$B = \frac{1 + P_0}{2}(f^0)^2 - Q_0 f^0(x - y)\cosh^{-1} U - \frac{1 - P_0}{2}N,$$

$$C = \frac{1 + P_0'}{2}(f^0)^2 - Q_0' f^0(x - y)\cosh^{-1} U - \frac{1 - P_0'}{2}N,$$

$$K_+ = \alpha(x - y)^{1 + 2(1 - \gamma - \delta)} \cdot \frac{1 - P_0}{2}(x + 1)^{-2\delta + 1}(1 - y)^{-2\gamma + 1},$$

$$K_- = \alpha(x - y)^{1 + 2(1 - \gamma - \delta)} \cdot \frac{1 + P_0}{2}(x - 1)^{-2\delta + 1}(1 + y)^{-2\gamma + 1},$$

$$L = \alpha^2 Q_0(x^2 - 1)^{-2\delta + 1}(1 - y^2)^{-2\gamma + 1}(x - y),$$

$$M = (x - y)^{4(1 - \gamma - \delta)} - \alpha^2(x^2 - 1)^{-2\delta + 1}(1 - y^2)^{-2\gamma + 1},$$

$$N = (x - y)^2 + (xy - 1)^2 + 2(x - y)(xy - 1) \cdot \tanh U.$$

Thus, these expressions determine a large class of solutions of the field equations.

Among stationary axisymmetric vacuum solutions of the Einstein equations, the most interesting are the asymptotically flat solutions reducing to the Zipoy solution in a static limit. For this case $\gamma = 0$, we have the expressions

$$f^0 = (x+1)^\delta (x-1)^{1-\delta}(1+y), \tag{3.20}$$

$$\lim_{U \to \infty} \tilde{f} = \left(\frac{x-1}{x+1}\right)^\delta. \tag{3.21}$$

In that case, to obtain the functions f and ω, it is sufficient to set in (3.19) $\gamma = 0$.

Consider two particular solutions arising from these relations.

1-I. Choosing

$$\delta = 1, \quad \gamma = 0, \quad a = d = p - 1,$$

$$b = c = q, \quad P_0 = 1/p, \quad Q_0 = -q/p,$$

$$P_0' = 0, \quad Q_0' = p/q, \quad R_0' = -1/q$$

in (3.19), one comes to the Kerr solution

$$f = \frac{p^2 x^2 + q^2 y^2 - 1}{(px+1)^2 + q^2 y^2}, \quad \omega = \frac{2k_0(1-y^2)(px+1)}{p(p^2 x^2 + q^2 y^2 - 1)},$$

where $p^2 + q^2 = 1$. When $q = 0$, the Kerr metric goes over into that of Schwarzschild.

Assumes its "standard" form with the transformations $k_0 x = r - m, y = \cos\theta, mp = k_0, mq = a, k_0^2 = m^2 - a^2$ yielding

$$f = \frac{r^2 - 2mr + a^2 \cos^2\theta}{r^2 + a^2 \cos^2\theta}, \quad \omega = \frac{2amr \sin^2\theta}{r^2 - 2mr + a^2 \cos^2\theta},$$

where the parameters m and a represent respectively the total mass and the angular momentum $J = ma$. When $a = 0$, the Kerr metric goes into that of Schwarzschild.

1-II. It would be interesting to find other asymptotically flat stationary generalizations of the Schwarzschild solution which, similar to the Kerr solution, would also have two physical parameters.

Setting $\delta = -1, d = 0$ in (3.19), we arrive at the solution

$$f = \tilde{f} = \frac{x+1}{x-1}\frac{\tilde{A}}{\tilde{B}}, \quad \Phi = \tilde{\Phi} = \frac{2\alpha\tilde{C}}{\tilde{B}}, \quad \omega = \tilde{\omega} = \frac{2k_0\alpha\tilde{D}}{\tilde{A}}, \qquad (3.22)$$

where

$$\tilde{A} = (x-y)^8 - \alpha^2(x^2-1)^3(1-y^2),$$
$$\tilde{B} = (x-y)^8 + \alpha^2(x-1)^4(x+1)^2(1+y)^2,$$
$$\tilde{C} = (x-y)^5(x+1)^2,$$
$$\tilde{D} = (x-y)^5(x-1)^3(1+y).$$

With the coordinate transformation $k_0 x = r - m, y = \cos\theta, k_0 = m \cdot S(\alpha)$, the asymptotic behavior of the metric coefficient f and ω takes the form

$$f \approx 1 - \frac{2m}{r} + O\left(\frac{1}{r^3}\right), \quad \omega \approx T(\alpha)\cdot\frac{\sin^2\theta}{r} + O\left(\frac{1}{r^2}\right),$$

from which follows that the angular momentum J is given by $J = -T'(\alpha)$, whereas m is the total mass. Therefore, the expression (3.22) define an asymptotically flat solution which reduces to the Schwarzschild metric by $\alpha = 0$.

2-I. If we choose

$$f^0 - 1, \quad \Phi^0 - \omega^0 - 0, \qquad (3.23)$$

in this case, we obtain from (3.8), (3.9) one-stationary Euclidon solution

$$f = f^0_{1,-} = k_0(xy-1) + k_0(x-y)\tanh U_0, \quad \Phi = \Phi^0_{1,-} = \frac{k_0(x-y)}{\cosh U_0},$$

$$\omega = \omega_{1,-} = \frac{(x-y)}{xy-1+(x-y)\tanh U_0}\frac{1}{\cosh U_0}, \qquad (3.24)$$

where $U = U_{1,-} = U_0$.

2-II. Let us consider

$$f^0 = f^0_{1,-}, \qquad \Phi^0 = \Phi^0_{1,-}, \tag{3.25}$$

in this case, we obtain from (3.8)–(3.11),

$$f = f^0_{2,-,-} = \frac{k_o(xy-1) + k_o(x-y)\tanh U}{f^0_{1,-}},$$

$$\Phi = \Phi^0_{2,-,-} = \frac{k_0}{f^0_{1,-}} \frac{(x-y)}{\cosh U} + \omega^0_{1,-},$$

where $U = U_{2,-,-} = \ln \frac{a}{b} = \text{const}$,

$$\frac{a}{b} = -\sqrt{\frac{1+\tanh U_0}{1-\tanh U_0}}, \quad \tanh U = \tanh U_0, \quad \cosh U = -\cosh U_0 \tag{3.26}$$

and

$$f^0_{2,-,-} = 1, \qquad \Phi^0_{2,-,-} = \omega^0_{2,-,-} = 0. \tag{3.27}$$

2-III. Two-stationary Euclidon solution.

If we choose the external field in the form

$$f^0 = f^0_{1,+} = k_0(xy+1) + k_0(x+y)\tanh U_0,$$

$$\Phi^0 = \Phi^0_{1,+} = \frac{k_0(x+y)}{\cosh U_0}, \tag{3.28}$$

$$\omega^0 = \omega^0_{1,+} = \frac{(x+y)}{xy+1+(x+y)\cdot\tanh U_0} \cdot \frac{1}{\cosh U_0},$$

where U_0 is a constant, we find [110]

$$\hat{L}_1 f^0 = k_0(1 + \tanh U_0), \quad \hat{L}_2 \Phi^0 = \frac{k_0}{\cosh U_0},$$

$$\hat{L}_2 f^0 = k_0(\tanh U_0 - 1), \quad \hat{L}_1 \Phi^0 = \frac{k_0}{\cosh U_0}, \tag{3.29}$$

$$k_0^{-1}a = (x+1)\sqrt{1+\tanh U_0} - \alpha_0(1-y)\sqrt{1-\tanh U_0},$$

$$k_0^{-1}b = \alpha_0(1+y)\sqrt{1+\tanh U_0} - (x-1)\sqrt{1-\tanh U_0}, \tag{3.30}$$

where α_0 is an integration constant. The function U in this case has the form

$$U = U_{2,-,+} = \ln[(x+1)\sqrt{1+\tanh U_0} - \alpha_0(1-y)\sqrt{1-\tanh U_0}]$$

$$- \ln[\alpha_0(1+y)\sqrt{1+\tanh U_0} - (x-1)\sqrt{1-\tanh U_0}]. \quad (3.31)$$

If we set

$$\tanh U_0 = \tanh U_0^a = \frac{1-a_0^2}{1+a_0^2}, \quad \alpha_0 = \frac{\varepsilon_2(b_0 - a_0)}{1+a_0 b_0} = \varepsilon_2 M_1, \quad (3.32)$$

where $M_1 = \frac{q_0}{p_0}$ ($\varepsilon_2 = \pm 1$) from (2.93), then, accordingly, a stationary two-Euclidean solution is constructed, which coincides with the Kerr–NUT solution in the form

$$f = f^0_{2,-,+} = \frac{k_o(xy-1) + k_o(x-y)\tanh U}{k_o(xy+1) + k_o(x+y)\tanh U_0} = \frac{A_2}{B_2},$$

$$\Phi = \Phi^0_{2,-,+} = \varepsilon_2 \frac{C_2}{B_2}, \quad \varepsilon_2 = 1,$$

$$A_2 = (x^2 - 1)(1 + a_0 b_0)^2 + (y^2 - 1)(b_0 - a_0)^2,$$

$$B_2 = [(1 + a_0 b_0)x + (1 - a_0 b_0)]^2 + [(b_0 - a_0)y + (b_0 + a_0)]^2,$$

$$C_2 = 2[(b_0 + a_0)(1 + a_0 b_0)x - (b_0 - a_0)(1 - a_0 b_0)y]. \quad (3.33)$$

If we set $a_0 = -b_0$, then (3.33) yields the Kerr solution.

2-IV. Let us consider also

$$f = f^0_{2,+,-} = \frac{k_o(xy+1) + k_o(x+y)\tanh U}{f^0_{1,-}},$$

$$\Phi = \Phi^0_{2,+,-} = \frac{k_0}{f^0_{1,-}} \frac{(x+y)}{\cosh U} + \omega^0_{1,-}, \quad (3.34)$$

where

$$U = U_{2,+,-}(x,y) = U_{2,-,+}(-x,y). \quad (3.35)$$

(b) Let us consider the seed metric (f^0, Φ^0, ω^0) in the Weyl canonical coordinates of the form

$$f^0_{1,z_2} = (z - z_2) + \sqrt{\rho^2 + (z - z_2)^2}\,\tanh U_0,$$

$$\Phi^0_{1,z_2} = \frac{\sqrt{\rho^2 + (z - z_2)^2}}{\cosh U_0},$$

$$\omega^0_{1.z_2} = \frac{\sqrt{\rho^2 + (z - z_2)^2}}{(z - z_2) + \sqrt{\rho^2 + (z - z_2)^2} \cdot \tanh U_0} \cdot \frac{1}{\cosh U_0}. \qquad (3.36)$$

The two-stationary Euclidon solution in the Weyl canonical coordinates using (3.4) in this case has the form

$$f = f^0_{2,z_1,z_2} = \frac{(z - z_1) + \sqrt{\rho^2 + (z - z_1)^2}\,\tanh U(\rho, z)}{f^0_{1,z_2}},$$

$$\Phi = \Phi^0_{2,z_1,z_2} = \frac{\sqrt{\rho^2 + (z - z_1)^2}}{f^0_{1,z_2} \cdot \cosh U(\rho, z)} + \omega^0_{1,z_2}(\rho, z), \qquad (3.37)$$

$$\omega = \omega^0_{2,z_1,z_2} = \frac{f^0_{1,z_2}\sqrt{\rho^2 + (z - z_1)^2}}{(z - z_1) + \sqrt{\rho^2 + (z - z_1)^2}\,\tanh U(\rho, z)} \cdot \frac{1}{\cosh U(\rho, z)}$$
$$+ \Phi^0_{1,z_2}(\rho, z).$$

The function U from (3.5) in this case has the form

$$U(\rho, z)$$
$$= U_{2,z_1,z_2} = \ln[(\sqrt{\rho^2 + (z - z_1)^2} + \sqrt{\rho^2 + (z - z_2)^2} + z_1 - z_2)$$
$$\times \sqrt{1 + \tanh U_0} - \alpha_0(z_1 - z_2 - \sqrt{\rho^2 + (z - z_2)^2}$$
$$+ \sqrt{\rho^2 + (z - z_1)^2})\sqrt{1 - \tanh U_0}] - \ln[\alpha_0(\sqrt{\rho^2 + (z - z_2)^2}$$
$$- \sqrt{\rho^2 + (z - z_1)^2} + z_1 - z_2)\sqrt{1 + \tanh U_0}$$
$$+ (z_1 - z_2 - \sqrt{\rho^2 + (z - z_2)^2} - \sqrt{\rho^2 + (z - z_1)^2})\sqrt{1 - \tanh U_0}],$$
$$\qquad (3.38)$$

where α_0 is an integration constant.

We can easily arrive at (3.38) from (3.31) with a shift

$$z \rightarrow z - \frac{z_1 + z_2}{2}, \qquad k_0 = \frac{z_1 - z_2}{2}$$

so that $z + k_0 \rightarrow z - z_2$, $z - k_0 \rightarrow z - z_1$.

If we put $z_1 = -z_2 = k_0$, then (3.37), (3.38) reduce to the Kerr–NUT solution (3.33).

3.4 Soliton in an Arbitrary Stationary Axially Symmetric Gravitational Field

As an another simplest solution of equations (3.1), the one-stationary soliton solution

$$f_s = C_1 \cdot \frac{\rho}{\rho_1 \coth U_0 + (z - z_1)},$$

$$\Phi_s = C_1 \cdot \frac{\rho_1}{\rho_1 \coth U_0 + (z - z_1)} \cdot \frac{1}{\sinh U_0} + C_2, \qquad (3.39)$$

$$\omega_s = \frac{1}{C_1} \cdot \frac{\rho_1}{\sinh U_0} + C_3,$$

where $\rho_1 = \sqrt{\rho^2 + (z - z_1)^2}$ and C_1, C_2, C_3, z_1 and U_0 are arbitrary constants, gives us again the possibility to obtain new solutions of (3.1) with the help of the method of again variation of parameters.

In this case, we have

$$f = \frac{\rho^2}{f_0(\rho, z)[\rho_1 \coth U(\rho, z) + (z - z_1)]},$$

$$\Phi = \frac{\rho_1 \cdot \rho}{f_0(\rho, z)[\rho_1 \coth U(\rho, z) + (z - z_1)]} \cdot \frac{1}{\sinh U(\rho, z)} + \omega_0(\rho, z),$$

$$\omega = \frac{f_0(\rho, z) \cdot \rho_1}{\rho \sinh U(\rho, z)} + \Phi_0(\rho, z), \qquad (3.40)$$

where $(f_0(\rho, z), \Phi_0(\rho, z), \omega_0(\rho, z))$ is an arbitrary stationary axially symmetric gravitational field and the unknown function $U(\rho, z)$ satisfies the equations

$$\rho_1 \frac{\partial U}{\partial \rho} = \frac{(z - z_1)}{\rho} f_0 \frac{\partial}{\partial \rho} \left(\frac{\rho}{f_0} \right) + f_0 \frac{\partial}{\partial z} \left(\frac{\rho}{f_0} \right) + \cosh U \cdot \frac{\rho}{f_0} \frac{\partial \Phi_0}{\partial \rho},$$

$$- [\rho_1 \sinh U + (z - z_1) \cosh U] \frac{1}{f_0} \frac{\partial \Phi_0}{\partial z}, \qquad (3.41)$$

$$\rho_1 \frac{\partial U}{\partial z} = -f_0 \frac{\partial}{\partial \rho} \left(\frac{\rho}{f_0} \right) + \frac{(z - z_1)}{\rho} f_0 \frac{\partial}{\partial z} \left(\frac{\rho}{f_0} \right)$$

$$+ [\rho_1 \sinh U + (z - z_1) \cosh U] \frac{1}{f_0} \frac{\partial \Phi_0}{\partial \rho} + \cosh U \cdot \frac{\rho}{f_0} \frac{\partial \Phi_0}{\partial z}.$$

Integration of (3.41) in general case is a difficult problem.

So, we turn to the examination of the superposition formulae corresponding to $\Phi_0 = \omega_0 = 0$, i.e., we choose as "seed" metric an arbitrary vacuum Weyl solution. It is easy to see that in this case the soliton method

$((3.40), (3.41))$ and the Euclidon method $((3.4), (3.5))$ coincide if we put $f^0 = \frac{\rho}{f_0}$.

In this case, in the prolate ellipsoidal coordinates (x, y), we have

$$\tilde{f} = \frac{1}{f_0}\frac{k_0(x^2 - 1)(1 - y^2)}{(xy - 1) + (x - y)\coth U}, \quad k_0 = \text{const.},$$

$$\tilde{\Phi} = \frac{1}{f_0}\frac{\sqrt{(x^2 - 1)(1 - y^2)}(x - y)}{(xy - 1) + (x - y)\coth U} \cdot \frac{1}{\sinh U},$$

$$\tilde{\omega} = f_0\frac{(x - y)}{\sqrt{(x^2 - 1)(1 - y^2)}} \cdot \frac{1}{\sinh U},$$

$$\frac{\partial U}{\partial x} = -\frac{1}{x^2 - 1} - \frac{(xy - 1)}{(x - y)} \cdot \frac{1}{f_0}\frac{\partial f_0}{\partial x} - \frac{(1 - y^2)}{(x - y)} \cdot \frac{1}{f_0}\frac{\partial f_0}{\partial y},$$

$$\frac{\partial U}{\partial y} = -\frac{1}{1 - y^2} + \frac{x^2 - 1}{(x - y)} \cdot \frac{1}{f_0}\frac{\partial f_0}{\partial x} - \frac{(xy - 1)}{(x - y)} \cdot \frac{1}{f_0}\frac{\partial f_0}{\partial y}. \tag{3.42}$$

Let us consider the physically most interesting solutions arising from (3.42).

If we choose

$$f_0 = e^{-2\psi}(x + 1)^{1+\delta}(x - 1)^{-\delta}(1 - y),$$

$$\psi = -\frac{c_0 x + d_0 y}{x^2 - y^2},$$

where δ, c_0 and d_0 are arbitrary real constants, we obtain

$$U = \ln[(x - y)^{2(1+\delta)} \cdot (x^2 - 1)^{-(\delta+1/2)} \cdot (1 - y^2)^{-1/2}]$$

$$+ \frac{y(d_0 - c_0)}{x + y} - \frac{(xy - 1)(d_0 + c_0)}{(x - y)^2} - \ln\beta,$$

$$\tilde{f} = \frac{e^{2\psi}}{(x + 1)^{1+\delta}(x - 1)^{-\delta}(1 - y)} \cdot \frac{k_0(x^2 - 1)(1 - y^2)}{[(xy - 1) + (x - y)\coth U]},$$

$$\tilde{\Phi} = \frac{e^{2\psi}}{(x + 1)^{1+\delta}(x - 1)^{-\delta}(1 - y)} \cdot \frac{k_0\sqrt{(x^2 - 1)(1 - y^2)}}{[(xy - 1) + (x - y)\coth U]} \cdot \frac{(x - y)}{\sinh U},$$

$$\tilde{\omega} = \frac{e^{-2\psi}(x + 1)^{1+\delta}(x - 1)^{-\delta}(1 - y)}{\sqrt{(x^2 - 1)(1 - y^2)}} \cdot \frac{(x - y)}{\sinh U}.$$

In a static limit $(\beta \to 0, U \to \infty, \sinh U \to \infty, \coth U \to 1)$, we have

$$\Phi = \tilde{\omega} = 0, \quad \tilde{f} = e^{2\psi} \cdot \left(\frac{x - 1}{x + 1}\right)^{\delta}.$$

Using the transformational theorem, we can construct a new class of solutions

$$f = \frac{R_0'\tilde{f}}{\frac{1+P_0}{2} - Q_0\tilde{\Phi} - \frac{1+P_0}{2}(\tilde{f}^2 + \tilde{\Phi}^2)},$$

$$\Phi = \frac{\frac{1+P_0'}{2} - Q_0'\tilde{\Phi} - \frac{1-P_0'}{2}(\tilde{f}^2 + \tilde{\Phi}^2)}{\frac{1+P_0}{2} - Q_0\tilde{\Phi} - \frac{1-P_0}{2}(\tilde{f}^2 + \tilde{\Phi}^2)},$$

$$(3.43)$$

where P_0, Q_0, P_0', Q_0' and R_0' are constants $(P_0^2 - Q_0^2 = 1, P_0'^2 + R_0'^2 = Q_0'^2 + 1)$.

If we consider the seed metric $(\frac{\rho}{f_0}, \omega_0 = 0)$, where f_0 is the static soliton solution, using Ehlers transformations (2.28), we obtain the Kerr solution [109].

In the case when the seed metric is the stationary soliton solution, the two-stationary soliton solution [56] can be obtained which also coincides with the Kerr–NUT solution.

3.5   Superposition of Two-Stationary Euclidon Solution with an Arbitrary Stationary Vacuum Field

1. Nonlinear superposition of the Kerr solution with an arbitrary axially symmetric Einstein field can be fulfilled by the variation of constants in the solution obtainable from the Kerr one by the constant phase transformation. We carry out the consideration in terms of the functions f and Φ satisfying, as is known, the equations

$$f\Delta f = (\vec{\nabla} f)^2 - (\vec{\nabla}\Phi)^2, \quad f\Delta\Phi = 2\vec{\nabla}f\,\vec{\nabla}\Phi. \qquad (3.44)$$

In the case of the superposition of the Kerr solution with an arbitrary vacuum Weyl field $f^0 = \exp(2\psi)$, it is sufficient to variate a single constant; as a "seed" variating solution, one can take the following:

$$f = \frac{x^2 - 1 + a_0^2(y^2 - 1)}{(x+1)^2 + a_0^2(y-1)^2}, \quad \Phi = \frac{2a_0(x+y)}{(x+1)^2 + a_0^2(y-1)^2}, \qquad (3.45)$$

where a_0 is an arbitrary real constant. Let us rewrite (3.45) in the form

$$\frac{f-1}{1} = -\frac{2[x+1-a_0^2(y-1)]}{(x+1)^2 + a_0^2(y-1)^2}, \quad \frac{\Phi-0}{1} = \frac{2a_0(x+y)}{(x+1)^2 + a_0^2(y-1)^2}.$$

$$(3.46)$$

Then, the superposition of the solution (3.46) with an arbitrary Weyl solution f^0 is written in the following way:

$$\frac{f - f^0}{f^0} = -\frac{2[x + 1 - a^2(y - 1)]}{(x + 1)^2 + a^2(y - 1)^2}, \quad \frac{\Phi - 0}{f^0} = \frac{2a(x + y)}{(x + 1)^2 + a^2(y - 1)^2}.$$
$$(3.47)$$

Here, the function f^0 satisfies the equation $f^0 \Delta f^0 = (\vec{\nabla} f^0)^2$ and the function a, as follows from (3.47), obeys the differential relations

$$(x + y)f^0 a_{,x} = -a[(xy + 1)f^0{}_{,x} + (1 - y^2)f^0{}_{,y}],$$
$$(x + y)f^0 a_{,y} = -a[-(x^2 - 1)f^0{}_{,x} + (xy + 1)f^0{}_{,y}]. \qquad (3.48)$$

2. To perform the superposition of the two-stationary Kerr solution with an arbitrary solution of system (3.44), it is not enough to variate one constant because both the functions f^0 and Φ^0 are not equal to zero. Therefore, we should take as "seed" variating solution not solution (3.46) but its generalization (3.33) containing two independent constants a_0 and b_0, e.g., the solution of the form

$$\frac{f - 1}{1} = -\frac{2[x(1 - a_0^2 b_0^2) + y(b_0^2 - a_0^2) + (a_0 + b_0)^2 + (a_0 b_0 - 1)^2]}{[x(1 + a_0 b_0) + 1 - a_0 b_0]^2 + [y(b_0 - a_0) + a_0 + b_0]^2},$$
$$\frac{\Phi - 0}{1} = \frac{2[x(a_0 + b_0)(1 + a_0 b_0) + y(a_0 - b_0)(1 - a_0 b_0)]}{[x(1 + a_0 b_0) + 1 - a_0 b_0]^2 + [y(b_0 - a_0) + a_0 + b_0]^2}, \qquad (3.49)$$

which reduces to (3.46) at $b_0 = 0$ and is just the Kerr solution at $b_0 = -a_0$.

Then, the superposition formulae generalizing in the case $\Phi^0 \neq 0$ takes the form

$$\frac{f - f^0}{f^0} = -\frac{2[x(1 - a^2 b^2) + y(b^2 - a^2) + (a + b)^2 + (1 - ab)^2]}{[x(1 + ab) + 1 - ab]^2 + [y(b - a) + a + b]^2},$$
$$\frac{\Phi - \Phi^0}{f^0} = \frac{2[x(a + b)(1 + ab) + y(a - b)(1 - ab)]}{[x(1 + ab) + 1 - ab]^2 + [y(b - a) + a + b]^2}, \qquad (3.50)$$

where a and b are some functions of the coordinates (x, y) and should be determined from the field equations.

Substituting (3.50) into (3.44), we can find the dependence of a and b on the functions f^0 and Φ^0:

$$a_{,x} = -\frac{1}{2(x + y)f^0}\{[(xy + 1)f^0{}_{,x} + (1 - y^2)f^0{}_{,y}]2a$$
$$+ [(xy + 1)\Phi^0{}_{,x} + (1 - y^2)\Phi^0{}_{,y}](1 - a^2) + (x + y)(1 + a^2)\Phi^0{}_{,x}\},$$

$$a_{,y} = -\frac{1}{2(x+y)f^0}\{[-(x^2-1)f^0{}_{,x} +(xy+1)f^0{}_{,y}]2a$$
$$+ [-(x^2-1)\Phi^0{}_{,x} +(xy+1)\Phi^0{}_{,y}](1-a^2) + (x+y)(1+a^2)\Phi^0{}_{,y}\},$$
$$(3.51)$$

$$b_{,x} = \frac{1}{2(x-y)f^0}\{[(xy-1)f^0{}_{,x} +(1-y^2)f^0{}_{,y}]2b$$
$$+ [(xy-1)\Phi^0{}_{,x} +(1-y^2)\Phi^0{}_{,y}](1-b^2) - (x-y)(1+b^2)\Phi^0{}_{,x}\},$$

$$b_{,y} = \frac{1}{2(x-y)f^0}\{[-(x^2-1)f^0{}_{,x} +(xy-1)f^0{}_{,y}]2b$$
$$+ [-(x^2-1)\Phi^0{}_{,x} +(xy-1)\Phi^0{}_{,y}](1-b^2) - (x-y)(1+b^2)\Phi^0{}_{,y}\}.$$

Relations (3.50) and (3.51) determine the desired superposition of the solution (3.49) with an arbitrary Einstein vacuum field.

If we put [112]

$$a = \exp(-U^a), \quad b = -\exp U^b,$$

we obtain from (3.51)

$$\frac{\partial U^a}{\partial x} = \left(\frac{xy+1}{x+y}\right)\frac{1}{f^0}\frac{\partial f^0}{\partial x} + \left(\frac{1-y^2}{x+y}\right)\frac{1}{f^0}\frac{\partial f^0}{\partial y}$$
$$+ \left[\cosh U^a + \left(\frac{xy+1}{x+y}\right)\sinh U^a\right]\frac{1}{f^0}\frac{\partial \Phi_0}{\partial x} + \left(\frac{1-y^2}{x+y}\right)\frac{\sinh U^a}{f^0}\frac{\partial \Phi_0}{\partial y},$$

$$\frac{\partial U^a}{\partial y} = -\left(\frac{x^2-1}{x+y}\right)\frac{1}{f^0}\frac{\partial f^0}{\partial x} + \left(\frac{xy+1}{x+y}\right)\frac{1}{f^0}\frac{\partial f^0}{\partial y}$$
$$+ \left[\cosh U^a + \left(\frac{xy+1}{x+y}\right)\sinh U^a\right]\frac{1}{f^0}\frac{\partial \Phi_0}{\partial y} - \left(\frac{x^2-1}{x+y}\right)\frac{\sinh U^a}{f^0}\frac{\partial \Phi_0}{\partial x}$$
$$(3.52)$$

and

$$\frac{\partial U^b}{\partial x} = \left(\frac{xy-1}{x-y}\right)\frac{1}{f^0}\frac{\partial f^0}{\partial x} + \left(\frac{1-y^2}{x-y}\right)\frac{1}{f^0}\frac{\partial f^0}{\partial y}$$
$$+ \left[\cosh U^b + \left(\frac{xy-1}{x-y}\right)\sinh U^b\right]\frac{1}{f^0}\frac{\partial \Phi_0}{\partial x} + \left(\frac{1-y^2}{x-y}\right)\frac{\sinh U^b}{f^0}\frac{\partial \Phi_0}{\partial y},$$

$$\frac{\partial U^b}{\partial y} = -\left(\frac{x^2-1}{x-y}\right)\frac{1}{f^0}\frac{\partial f^0}{\partial x} + \left(\frac{xy-1}{x-y}\right)\frac{1}{f^0}\frac{\partial f^0}{\partial y}$$

$$+ \left[\cosh U^b + \left(\frac{xy-1}{x-y}\right)\sinh U^b\right]\frac{1}{f^0}\frac{\partial \Phi_0}{\partial y} - \left(\frac{x^2-1}{x-y}\right)\frac{\sinh U^b}{f^0}\frac{\partial \Phi_0}{\partial x}.$$

$$(3.53)$$

3. We show now that such solutions with required properties can be obtained from the relations (3.50) and (3.51). To do this, we first write down the superposition of the solution (3.49) with an arbitrary static vacuum field, which follows from the general case by putting $\Phi^0 = 0, f^0 = \exp\psi$:

$$f + i\Phi = \varepsilon = e^\psi\left[1 - \frac{2(1-ia)(1-ib)}{x(1+ab) + iy(a-b) + (1-ia)(1-ib)}\right], \quad (3.54)$$

ψ being an arbitrary solution of the equation $\Delta\psi = 0$, whereas, the functions a and b should be found for every given ψ from the equations

$$\begin{aligned}
a_{,x} &= -a(x+y)^{-1}\left[(xy+1)\psi_{,x} + (1-y^2)\psi_{,y}\right], \\
a_{,y} &= -a(x+y)^{-1}\left[-(x^2-1)\psi_{,x} + (xy+1)\psi_{,y}\right], \\
b_{,x} &= b(x-y)^{-1}\left[(xy-1)\psi_{,x} + (1-y^2)\psi_{,y}\right], \\
b_{,y} &= b(x-y)^{-1}\left[-(x^2-1)\psi_{,x} + (xy-1)\psi_{,y}\right].
\end{aligned} \qquad (3.55)$$

3-I. Choosing as ψ the Zipoy solution

$$\psi = \delta \ln \frac{x-1}{x+1}, \qquad (3.56)$$

one can find, integrating (3.55), that

$$a = \frac{\alpha(x+y)^{2\delta}}{(x^2-1)^\delta}, \qquad b = \frac{\beta(x-y)^{2\delta}}{(x^2-1)^\delta} \qquad (3.57)$$

(α and β are real constants).

Remembering that once ε is a solution of the Ernst equation, then $\tilde{\varepsilon} = 1/\varepsilon$ also satisfies this equation, and putting in (3.56), (3.57) $\delta = -2, \alpha = -\beta$, we come to the following stationary solution:

$$\tilde{\varepsilon} = \tilde{f} + i\tilde{\Phi} = \frac{x-1}{x+1}\cdot\frac{E_-}{E_+}, \qquad (3.58)$$

where

$$E_\pm = (x^2 - y^2)^4 - \alpha^2(x \pm 1)^2(x^2 - 1)^3$$

$$- 2i\alpha y(x \pm 1)(x^2 - 1)(x^4 + 6x^2 y^2 + y^4 \mp 4x^3 \mp 4xy^2),$$

$$\tilde{f} = \frac{x-1}{x+1} \cdot \frac{A}{B}, \quad \tilde{\Phi} = \frac{4\alpha y(x-1)^2 C}{B}, \quad \tilde{\omega} = -\frac{4k_0\alpha(1-y)^2 D}{(1-\alpha^2)A},$$

$$e^{2\gamma} = \frac{x^2 - 1}{x^2 - y^2} \cdot \frac{A}{(1-\alpha^2)^2(x^2 - y^2)^8}, \tag{3.59}$$

where

$$A \equiv [(x^2 - y^2)^4 - \alpha^2(x^2 - 1)^4]^2 + 4\alpha^2(x^2 - 1)^3(y^2 - 1)(x^4 + 6x^2 y^2 + y^4)^2,$$

$$B \equiv [(x^2 - y^2)^4 - \alpha^2(x+1)^2(x-1)^3]^2 + 4\alpha^2 y^2(x+1)^2(x^2 - 1)^2$$

$$\times (x^4 + 6x^2 y^2 + y^4 - 4x^3 - 4xy^2)^2,$$

$$C \equiv (x^2 - y^2)^4(-3x^4 + 2x^2 y^2 + y^4) + \alpha^2(x^2 - 1)^4(5x^4 + 10x^2 y^2 + y^4),$$

$$D \equiv (x^2 - y^2)^5(3x^5 + 3x^4 + x^3 + x^3 y^2 + 6x^2 y^2 + 3xy^2 - y^4)$$

$$- \alpha^2(x^2 - y^2)^5(x+1)^3(3x^2 - 3x + y^2 + 1)$$

$$- \alpha^2(x^2 - 1)^3(5x^9 + 8x^8 + 10x^7 + 10x^7 y^2 - 6x^6 y^2 + 5x^6 + 5x^6 y^2$$

$$+ x^5 + 76x^5 y^2 + x^5 y^4 + 45x^4 y^2 - 145x^4 y^4 + 10x^3 y^2 + 10x^3 y^4$$

$$+ 15x^2 y^4 - 51x^2 y^6 + 5xy^4 - y^6 - 3y^8) + \alpha^4(x^2 - 1)^3(x+1)^5$$

$$\times (5x^4 - 10x^3 + 10x^2 + 10x^2 y^2 - 5x - 5xy^2 + y^4 + y^2 + 1). \tag{3.60}$$

With the coordinate transformation

$$k_0 x = r - m, \quad y = \cos\theta, \quad k_0 = m(1 - \alpha^2)(1 - 3\alpha^2)^{-1}, \tag{3.61}$$

the asymptotic behavior of the metric coefficients $\tilde{f}, \tilde{\gamma}$ and $\tilde{\omega}$ takes the form

$$\tilde{f} = 1 - \frac{2m}{r} + o\left(\frac{1}{r^3}\right), \quad e^{2\tilde{\gamma}} = 1 + o\left(\frac{1}{r^2}\right),$$

$$\tilde{\omega} = -\frac{4m^2\alpha(3 - 5\alpha^2)\sin^2\theta}{(1 - 3\alpha^2)^2 r} + o\left(\frac{1}{r^2}\right), \tag{3.62}$$

from which follows that the angular momentum $\tilde{J}$ is given by

$$\tilde{J} = -\frac{2m^2\alpha(3 - 5\alpha^2)}{(1 - 3\alpha^2)^2}, \tag{3.63}$$

where m is the total mass.

Therefore, formulae (3.59), (3.60) define an asymptotically flat solution which reduces to the Schwarzschild metric by putting the rotation parameter α equal to zero.

3-II. We briefly consider some non-algebraic solutions representing a stationary rotating body. These solutions arise from the superposition formulae (3.54), (3.55) with the choice of ψ in the form of pure multipole moments, for instance, in the form

$$\psi = \alpha_0 \left(\alpha_n \tilde{L}_n \ln \frac{x-1}{x+1} \right), \quad \alpha_0, \alpha_n = \text{const}, \quad n = 1, 2, \ldots, \quad (3.64)$$

where

$$\tilde{L}_n = \{(x^2 - y^2)^{-1} \left[y(x^2 - 1)\partial_x + x(1 - y^2)\partial_y \right]\}^n \quad (3.65)$$

is the operator in which the expression enclosed in braces acts n times upon the function $\ln[(x-1)/(x+1)]$.

In a particular case $n = 1, \alpha = 1$, we get for a_1 and b from (3.55), (3.64) and (3.65)

$$a = -\alpha \exp \left\{ \frac{\alpha[x(1 - 3y^2) - y(1 + y^2)]}{2(x-y)(x+y)^2} \right\},$$

$$b = \alpha \exp \left\{ -\frac{\alpha[x(1 - 3y^2) + y(1 + y^2)]}{2(x+y)(x-y)^2} \right\}. \quad (3.66)$$

Thus, the asymptotically flat solutions defined by (3.54), (3.64)–(3.66) also reduce to the Schwarzschild solution by putting $\alpha = 0$ in the expressions for metric functions.

4. It turns out possible to generalize the solution considered above if we rewrite the superposition formulae (3.54), (3.55) in the Weyl canonical coordinates (ρ, z), i.e., in the form

$$\varepsilon = e^\psi \frac{r_+ A + r_- B - 2l}{r_+ A + r_- B + 2l}, \quad r_\pm \equiv [\rho^2 + (z \pm l)^2]^{1/2}, \quad (3.67)$$

where A and B are determined by the relations

$$A = \frac{1 + ia}{1 - ia}, \quad B = \frac{1 + ib}{1 - ib},$$

and the functions a and b should be found from the first-order differential equations:

$$r_+ a_{,\rho} = -a[(z + l)\psi_{,\rho} + \rho\psi_{,z}], \quad r_+ a_{,z} = -a[-\rho\psi_{,\rho} + (z + l)\psi_{,z}],$$

$$r_- b_{,\rho} = b[(z - l)\psi_{,\rho} + \rho\psi_{,z}], \quad r_- b_{,z} = b[-\rho\psi_{,\rho} + (z - l)\psi_{,z}]. \quad (3.68)$$

Note that in the limiting static case, (3.67) is reduced to (1.43).

To perform the required generalization, we choose ψ in the form

$$\psi = \ln\left(\frac{z - m + R_+}{z + m + R_+} \cdot \frac{z + l + r_+}{z - l + r_-}\right), \quad R_\pm \equiv \left[\rho^2 + (z \pm m)^2\right]^2, \quad (3.69)$$

where m is a real constant.

The integration of (3.68) then yields the expressions for a and b:

$$a = -\alpha\frac{\rho^2 + (z + l)(z + m) + r_+ R_+}{\rho^2 + (z + l)(z - m) + r_+ R_-} \cdot \frac{\rho^2 + z^2 - l^2 + r_- r_+}{2r_+^2},$$

$$b = \alpha\frac{\rho^2 + (z - l)(z - m) + r_- R_-}{\rho^2 + (z - l)(z + m) + r_- R_+} \cdot \frac{\rho^2 + z^2 - l^2 + r_- r_+}{2r_-^2}$$

$$(3.70)$$

with a special choice of the integration constants ensuring the asymptotic flatness of the solution.

The asymptotic behavior of the functions f and Φ defined by (3.64), (3.67)–(3.70) in the spherical coordinates (R, θ) ($\rho = R\sin\theta, z = R\cos\theta$) has the form

$$f = 1 - \frac{2M}{R} + O\left(\frac{1}{R^2}\right), \quad \Phi = \frac{2J\cos\theta}{R^2} + O\left(\frac{1}{R^3}\right),$$

M and J being the total mass and the angular momentum, respectively,

$$M = [(m - l)(1 - \alpha^2) + l(1 + \alpha^2)](1 - \alpha^2)^{-1},$$

$$J = 2\alpha l[2m(1 - \alpha^2) - l(1 - 3\alpha^2)](1 - \alpha^2)^{-2}.$$

Therefore, we obtained a stationary vacuum solution which reduces to the Schwarzschild metric at $\alpha = 0$.

5. It turns out possible to generalize two-stationary Euclidon solution in the Weyl canonical coordinates (3.37), (3.38) with an arbitrary stationary vacuum field considered above if we rewrite the superposition formulae (3.52), (3.53) in the Weyl canonical coordinates (ρ, z). For this purpose, it is better to use a single relation for a complex function $\varepsilon = f + i\Phi$, i.e.,

$$\varepsilon = \frac{(\varepsilon^0 - \varepsilon^{0*})}{2} + \frac{(\varepsilon^0 + \varepsilon^{0*})}{2} \cdot \frac{r_2 A + r_1 B + z_2 - z_1}{r_2 A + r_1 B + z_1 - z_2},$$

$$(3.71)$$

$$r_i \equiv \left[\rho^2 + (z - z_i)^2\right]^{1/2}; \quad i = 1, 2,$$

where $\varepsilon^0 = f^0 + i\Phi^0$ defines a known solution of the Ernst equation.

A and B are determined by the relations

$$A = \frac{1 + ia}{1 - ia}, \quad B = \frac{1 + ib}{1 - ib},$$

and the functions $a = \exp\left(-U^a\right)$, $b = -\exp U^b$ should be found from the first-order differential equations:

$$\sqrt{\rho^2 + (z - z_2)^2}\,\frac{\partial U^a}{\partial \rho}$$

$$= (z - z_2)\frac{1}{f^0}\frac{\partial f^0}{\partial \rho} + \frac{\rho}{f^0}\frac{\partial f^0}{\partial z} + \left[(z - z_2)\sinh U\right.$$

$$\left. + \sqrt{\rho^2 + (z - z_2)^2}\cdot\cosh U\right]\frac{1}{f^0}\frac{\partial \Phi^0}{\partial \rho} + \rho\frac{\sinh U}{f^0}\frac{\partial \Phi^0}{\partial z},$$

$$\sqrt{\rho^2 + (z - z_2)^2}\,\frac{\partial U^a}{\partial z}$$

$$= -\frac{\rho}{f^0}\frac{\partial f^0}{\partial \rho} + \frac{z - z_2}{f^0}\frac{\partial f^0}{\partial z} + \left[(z - z_2)\sinh U\right.$$

$$\left. + \sqrt{\rho^2 + (z - z_2)^2}\cdot\cosh U\right]\frac{1}{f^0}\frac{\partial \Phi^0}{\partial z} - \rho\frac{\sinh U}{f^0}\frac{\partial \Phi^0}{\partial \rho} \qquad (3.72)$$

and

$$\sqrt{\rho^2 + (z - z_1)^2}\,\frac{\partial U^b}{\partial \rho}$$

$$= (z - z_1)\frac{1}{f^0}\frac{\partial f^0}{\partial \rho} + \frac{\rho}{f^0}\frac{\partial f^0}{\partial z} + \left[(z - z_1)\sinh U\right.$$

$$\left. + \sqrt{\rho^2 + (z - z_1)^2}\cdot\cosh U\right]\frac{1}{f^0}\frac{\partial \Phi^0}{\partial \rho} + \rho\frac{\sinh U}{f^0}\frac{\partial \Phi^0}{\partial z},$$

$$\sqrt{\rho^2 + (z - z_1)^2}\,\frac{\partial U^b}{\partial z}$$

$$= -\frac{\rho}{f^0}\frac{\partial f^0}{\partial \rho} + \frac{z - z_1}{f^0}\frac{\partial f^0}{\partial z} + \left[(z - z_1)\sinh U\right.$$

$$\left. + \sqrt{\rho^2 + (z - z_1)^2}\cdot\cosh U\right]\frac{1}{f^0}\frac{\partial \Phi^0}{\partial z} - \rho\frac{\sinh U}{f^0}\frac{\partial \Phi^0}{\partial \rho}. \qquad (3.73)$$

3.6   Three- and Four-Stationary Euclidon Solutions

1. The following formulas for three- and four-stationary Euclidon solutions in the prolate ellipsoidal coordinates were obtained in Ref. [112]:

1-I. Let us consider superposition (3.50) of the stationary two-Euclidon solution (3.49) with the stationary own-Euclidon solution $(f^0_{1,-}, \ \ \Phi^0_{1,-})$ (3.24).

This stationary three-Euclidon solution is easy to rewrite as

$$f = f^0_{3,-,+,-} = \frac{k_0(xy-1) + k_0(x-y)\tanh U(a,b)}{f^0_{2,+,-}},$$

$$\Phi = \Phi^0_{3,-,+,-} = \frac{1}{f^0_{2,+,-}} \cdot \frac{k_0(x-y)}{\cosh U(a,b)} + \omega^0_{2,+,-}, \tag{3.74}$$

where

$$f^0_{2,+,-} = \frac{k_0(xy+1) + k_0(x+y)\tanh U^a}{f^0_{1,-}},$$

$$\omega^0_{2,+,-} = f^0_{1,-} \frac{k_0(x+y)}{k_0(xy+1) + k_0(x+y)\tanh U^a} \cdot \frac{1}{\cosh U^a} + \Phi^0_{1,-}. \tag{3.75}$$

And from (3.31), (3.52), (3.53), (3.26) and (3.34),

$$U(a,b) = U_{3,-,+,-}(x,y)$$
$$= \ln[(x+1)\sqrt{1+\tanh U^a} - \alpha(1-y)\sqrt{1-\tanh U^a}]$$
$$- \ln[\alpha(1+y)\sqrt{1+\tanh U^a} - (x-1)\sqrt{1-\tanh U^a}], \tag{3.76}$$

where

$$\tanh U^a = \frac{1-a^2}{1+a^2}, \quad \alpha = \varepsilon_3 \frac{b-a}{1+ab}, \quad \varepsilon_3 = \pm 1,$$

$$a = \exp(-U^a), \quad b = -\exp U^b, \tag{3.77}$$

$$U^a = U_{2,+,-}(x,y), \quad U^b = U_{2,-,-}(x,y).$$

From (3.74), it is easy to see that these formulas also define a superposition of the stationary own-Euclidon solution (3.8) with the stationary two-Euclidon solution $(f^0_{2,+,-}, \ \ \omega^0_{2,+,-})$ with $U = U(a,b) = U_{3,-,+,-}(x,y)$.

1-II. Let us consider also

$$f = f^0_{3,-,-,+} = \frac{k_0(xy-1) + k_0(x-y)\tanh U(a,b)}{f^0_{2,-,+}},$$

$$\Phi = \Phi^0_{3,-,-,+} = \frac{1}{f^0_{2,-,+}} \cdot \frac{k_0(x-y)}{\cosh U(a,b)} + \omega^0_{2,-,+}, \tag{3.78}$$

where

$$f^0_{2,-,+} = \frac{k_0(xy-1) + k_0(x-y)\tanh U^a}{f^0_{1,+}},$$

$$\omega^0_{2,-,+} = f^0_{1,+}\frac{(x-y)}{xy-1+(x-y)\cdot\tanh U^a} \cdot \frac{1}{\cosh U^a} + \Phi^0_{1,+}. \tag{3.79}$$

And from (3.28), (3.31), (3.52), (3.53), (3.26) and (3.34),

$$U(a,b) = U_{3,-,-,+}(x,y)$$
$$= \ln[(x+1)\sqrt{1+\tanh U^a} - \alpha(1-y)\sqrt{1-\tanh U^a}]$$
$$\quad - \ln[\alpha(1+y)\sqrt{1+\tanh U^a} - (x-1)\sqrt{1-\tanh U^a}], \tag{3.80}$$

where

$$\tanh U^a = \tanh U_0 = \frac{1-a^2}{1+a^2}, \quad \alpha = \varepsilon_3\frac{b-a}{1+ab}, \quad \varepsilon_3 = \pm 1,$$

$$U^a = U_{2,+,+}(x,y) = U_{2,-,-}(x,y), \quad U^b = U_{2,-,+}(x,y), \tag{3.81}$$

$$a = \exp(-U^a) = -\sqrt{\frac{1-\tanh U_0}{1+\tanh U_0}},$$

$$b = -\exp U^b = -\frac{(x+1)\sqrt{1+\tanh U_0} - \alpha_0(1-y)\sqrt{1-\tanh U_0}}{\alpha_0(1+y)\sqrt{1+\tanh U_0} - (x-1)\sqrt{1-\tanh U_0}}.$$

If $\alpha_0 \to \infty$, $\varepsilon_3 = -1$, we have

$$\alpha = -\frac{\sqrt{1-\tanh^2 U_0}}{y+\tanh U_0},$$

$$\exp U(a,b) = -\sqrt{\frac{1+\tanh U_0}{1-\tanh U_0}}, \quad \tanh U(a,b) = \tanh U_0$$

and

$$f^0_{3,-,-,+} \to f^0_{1,+},$$

$$\Phi^0_{3,-,-,+} \to \Phi^0_{1,+}. \tag{3.82}$$

Similarly for $\alpha_0 \to 0$.

1-III. Superposition of the stationary two-Euclidon solution (3.50) with the stationary two-Euclidon solution $(f^0_{2,-,+}, \; \Phi^0_{2,-,+})$ gives stationary four-Euclidon solution

$$f = f^0_{4,-,+,-,+} = \frac{k_0(xy-1) + k_0(x-y)\tanh U(a,b)}{f^0_{3,+,-,+}},$$

$$\Phi = \Phi^0_{4,-,+,-,+} = \frac{1}{f^0_{3,+,-,+}} \cdot \frac{k_0(x-y)}{\cosh U(a,b)} + \omega^0_{3,+,-,+}, \tag{3.83}$$

where

$$f^0_{3,+,-,+}(x,y) = -f^0_{3,-,+,-}(-x,y),$$

$$\omega^0_{3,+,-,+}(x,y) = -\omega^0_{3,-,+,-}(-x,y). \tag{3.84}$$

And

$$U(a,b) = U_{4,-,+,-,+}(x,y)$$
$$= \ln[(x+1)\sqrt{1+\tanh U^a} - \alpha(1-y)\sqrt{1-\tanh U^a}]$$
$$- \ln[\alpha(1+y)\sqrt{1+\tanh U^a} - (x-1)\sqrt{1-\tanh U^a}], \tag{3.85}$$

where

$$\tanh U^a = \frac{1-a^2}{1+a^2}, \quad \alpha = \varepsilon_4 \frac{b-a}{1+ab}, \quad \varepsilon_4 = \pm 1,$$

$$a = \exp(-U^a), \quad b = -\exp U^b,$$

$$U^a = U_{3,+,-,+}(x,y), \quad U^b = U_{3,-,-,+}(x,y).$$

From (3.83), it is easy to see that these formulas also define a superposition of the stationary own-Euclidon solution (3.8) with the stationary three-Euclidon solution $(f^0_{3,+,-,+}, \; \omega^0_{3,+,-,+})$ with $U = U(a,b) = U_{4,-,+,-,+}(x,y)$.

In a similar way, we can get the stationary 2N-Euclidon solution containing N-Kerr solution on the symmetry axis in which all gravitating centers coincide with each other and reducing to the Zipoy solution (1.43) in a static limit.

2-I. Let us consider superposition (3.71) of the stationary two-Euclidon solution (3.37, 3.38) with the stationary own-Euclidon solution $(f^0_{1,z_3}, \; \Phi^0_{1,z_3})$ (3.36) in the Weyl canonical coordinates (ρ, z).

This stationary three-Euclidon solution is easy to rewrite as

$$f = f^0_{3,z_1,z_2,z_3} = \frac{(z - z_1) + \sqrt{\rho^2 + (z - z_1)^2}\,\tanh U(\rho, z)}{f^0_{2,z_2,z_3}},$$

$$\Phi = \Phi^0_{3,z_1,z_2,z_3} = \frac{\sqrt{\rho^2 + (z - z_1)^2}}{f^0_{2,z_2,z_3} \cdot \cosh U(\rho, z)} + \omega^0_{2,z_2,z_3}(\rho, z),$$

$$\omega = \omega^0_{3,z_1,z_2,z_3} = \frac{f^0_{2,z_2,z_3}\sqrt{\rho^2 + (z - z_1)^2}}{(z - z_1) + \sqrt{\rho^2 + (z - z_1)^2}\,\tanh U(\rho, z)} \cdot \frac{1}{\cosh U(\rho, z)}$$

$$+ \Phi^0_{2,z_2,z_3}(\rho, z) \tag{3.86}$$

where

$$f^0_{2,z_2,z_3} = \frac{(z - z_2) + \sqrt{\rho^2 + (z - z_2)^2}\,\tanh U(\rho, z)}{f^0_{1,z_3}},$$

$$\omega^0_{2,z_2,z_3} = f^0_{1,z_3} \frac{\sqrt{\rho^2 + (z - z_2)^2}}{(z - z_2) + \sqrt{\rho^2 + (z - z_2)^2} \cdot \tanh U^a} \cdot \frac{1}{\cosh U^a} + \Phi^0_{1,z_3}. \tag{3.87}$$

And from (3.38), (3.72) and (3.73),

$$U(a, b) = U_{3,z_1,z_2,z_3}(\rho, z)$$

$$= \ln[(\sqrt{\rho^2 + (z - z_1)^2} + \sqrt{\rho^2 + (z - z_2)^2} + z_1 - z_2)\sqrt{1 + \tanh U^a}$$

$$- \alpha(z_1 - z_2 - \sqrt{\rho^2 + (z - z_2)^2} + \sqrt{\rho^2 + (z - z_1)^2})\sqrt{1 - \tanh U^a}]$$

$$- \ln[\alpha(\sqrt{\rho^2 + (z - z_2)^2} - \sqrt{\rho^2 + (z - z_1)^2} + z_1 - z_2)\sqrt{1 + \tanh U^a}$$

$$+ (z_1 - z_2 - \sqrt{\rho^2 + (z - z_2)^2} - \sqrt{\rho^2 + (z - z_1)^2})\sqrt{1 - \tanh U^a}], \tag{3.88}$$

where

$$\tanh U^a = \frac{1 - a^2}{1 + a^2}, \quad \alpha = \varepsilon_3 \frac{b - a}{1 + ab}, \quad \varepsilon_3 = \pm 1,$$

$$a = \exp(-U^a), \quad b = -\exp(U^b),$$

$$U^a = U_{2,z_2,z_3}(\rho, z), \quad U^b = U_{2,z_1,z_3}(\rho, z). \tag{3.89}$$

From (3.86), it is easy to see that these formulas also define a superposition of the stationary own-Euclidon solution (3.4) with the stationary two-Euclidon solution $(f^0_{2,z_2,z_3}, \omega^0_{2,z_2,z_3})$ with $U = U(a, b) = U_{3,z_1,z_2,z_3}(\rho, z)$.

If $z_1 = z_2$ and $\alpha_0 \to \infty, \varepsilon_3 = -1$ (or $\alpha_0 \to 0$) in $U^a = U^b = U_{2,z_1,z_3}(\rho, z)$, we have

$$f^0_{3,z_1,z_1,z_3} \to f^0_{1,z_3},$$

$$\Phi^0_{3,z_1,z_1,z_3} \to \Phi^0_{1,z_3}. \tag{3.90}$$

2-II. Superposition of the stationary two-Euclidon solution (3.37) with the stationary two-Euclidon solution $(f^0_{2,z_3,z_4}, \Phi^0_{2,z_3,z_4})$ gives stationary four-Euclidon solution in the Weyl canonical coordinates (ρ, z)

$$f = f^0_{4,z_1,z_2,z_3,z_4} = \frac{(z - z_1) + \sqrt{\rho^2 + (z - z_1)^2}\,\tanh U(\rho, z)}{f^0_{3,z_2,z_3,z_4}},$$

$$\Phi = \Phi^0_{4,z_1,z_2,z_3,z_4} = \frac{\sqrt{\rho^2 + (z - z_1)^2}}{f^0_{3,z_2,z_3.z_4} \cdot \cosh U(\rho, z)} + \omega^0_{3,z_2,z_3,z_4}(\rho, z) \tag{3.91}$$

and

$$U(a, b) = U_{4,z_1,z_2,z_3,z_4}(\rho, z)$$

$$= \ln[(\sqrt{\rho^2 + (z - z_1)^2} + \sqrt{\rho^2 + (z - z_2)^2} + z_1 - z_2)\sqrt{1 + \tanh U^a}$$

$$- \alpha(z_1 - z_2 - \sqrt{\rho^2 + (z - z_2)^2} + \sqrt{\rho^2 + (z - z_1)^2})\sqrt{1 - \tanh U^a}]$$

$$- \ln[\alpha(\sqrt{\rho^2 + (z - z_2)^2} - \sqrt{\rho^2 + (z - z_1)^2} + z_1 - z_2)\sqrt{1 + \tanh U^a}$$

$$+ (z_1 - z_2 - \sqrt{\rho^2 + (z - z_2)^2} - \sqrt{\rho^2 + (z - z_1)^2})\sqrt{1 - \tanh U^a}], \tag{3.92}$$

where

$$\tanh U^a = \frac{1 - a^2}{1 + a^2}, \quad \alpha = \varepsilon_4 \frac{b - a}{1 + ab}, \quad \varepsilon_4 = \pm 1,$$

$$a = \exp(-U^a), \quad b = -\exp U^b,$$

$$U^a = U_{3,z_2,z_3,z_4}(\rho, z), \quad U^b = U_{3,z_1,z_3,z_4}(\rho, z). \tag{3.93}$$

From (3.91), it is easy to see that these formulas also define a superposition of the stationary own-Euclidon solution (3.4) with the stationary three-Euclidon solution $(f^0_{3,z_2,z_3,z_4}, \omega^0_{3,z_2,z_3,z_4})$ with $U = U(a, b) = U_{4,z_1,z_2,z_3,z_4}(\rho, z)$.

In a similar way, we can get the stationary 2N-Euclidon solution containing N-Kerr solution on the symmetry axis in which all gravitating centers do not coincide with each other and reducing to the static 2N-Euclidon (soliton) solution (1.99) in a static limit.

3.7 Some Physical Interpretation of the Kerr Solution

1. The Kerr metric (2.96) in curvature coordinates (r, θ) has the form

$$
ds^2 = \frac{r^2 + a^2 \cos^2 \theta}{r^2 - 2mr + a^2} dr^2 + [r^2 + a^2 \cos^2 \theta]
$$

$$
\times \left[d\theta^2 + \frac{r^2 - 2mr + a^2}{r^2 - 2mr + a^2 \cos^2 \theta} \sin^2 \theta d\varphi^2 \right] - \left[1 - \frac{2mr}{r^2 + a^2 \cos^2 \theta} \right]
$$

$$
\times \left\{ dt - \frac{2amr \sin^2 \theta}{r^2 - 2mr + a^2 \cos^2 \theta} d\varphi \right\}^2
$$

$$
= \frac{r^2 + a^2 \cos^2 \theta}{r^2 - 2mr + a^2} dr^2 + [r^2 + a^2 \cos^2 \theta] d\theta^2
$$

$$
+ \left(r^2 + a^2 + \frac{2mra^2 \sin^2 \theta}{r^2 + a^2 \cos^2 \theta} \right) \sin^2 \theta d\varphi^2 - \left[1 - \frac{2mr}{r^2 + a^2 \cos^2 \theta} \right] dt^2
$$

$$
- \frac{4amr}{r^2 + a^2 \cos^2 \theta} d\varphi \, dt, \tag{3.94}
$$

where the metric coefficients

$$
f_{K.} = \frac{r^2 - 2mr + a^2}{r^2 + a^2 \cos^2 \theta},
$$

$$
\omega_{K.} = \frac{2amr \sin^2 \theta}{r^2 - 2mr + a^2 \cos^2 \theta}, \tag{3.95}
$$

$$
e^{\gamma_{K.}} = \frac{r^2 - 2mr + a^2}{(r - m)^2 - (m^2 - a^2) \cos^2 \theta}.
$$

In the case of a slight deviation from spherical symmetry $(a \ll 1)$, we obtain the Lense–Thirring metric

$$
ds^2 \approx ds^2_{\text{L.-T.}} = \frac{dr^2}{g_{00}} + r^2(d\theta^2 + \sin^2 \theta d\varphi^2) - \frac{4am}{r} \sin^2 \theta d\varphi d\theta - g_{00}c^2 dt^2,
$$

$$
\tag{3.96}
$$

where $g_{00} = 1 - \frac{2m}{r}$.

Let's consider a small neighborhood around a fixed value of $r = r_0$, $\theta = \theta_0$ and by using transformations

$$
\begin{cases}
dt_{\text{L.-T.}} \approx dt_{\text{Sch.}} + a \sin^2 \theta_0 \cdot d\varphi_{\text{Sch.}} \\
d\varphi_{\text{L.-T.}} \approx d\varphi_{\text{Sch.}} + \omega_0 dt_{\text{Sch.}},
\end{cases}
\tag{3.97}
$$

where $\omega_0 = \frac{M}{mr_0^2}$ and $M = mac$, we can obtain the Schwarzschild metric (1.100)

$$ds_{\text{L.-T.}}^2 \approx ds_{\text{Sch.}}^2 = \frac{dr^2}{g_{00}} + r^2(d\theta^2 + \sin^2\theta \, d\varphi^2) - g_{00}c^2 dt^2. \qquad (3.98)$$

Using the formulas from section (1.10), item 1, you can arrive at geodesic motion in the equatorial plane ($\theta_0 = \frac{\pi}{2}$).

2. The stationary Euclidon solution has a clear physical interpretation as a relativistic accelerated non-inertial reference frame, which provides a different perspective on the physical interpretation of well-known solutions, such as the Kerr solution.

It is easy to see that the Kerr metric (2.1), (3.33), (3.95) is approximately equal to the metric of the stationary Euclidon solution (2.58), (2.59) in a small region of spacetime $\rho \approx \rho_0$, $z \approx z_0$ if we assume that

$$\begin{cases} C_1 = z_0 - z_2 + \sqrt{\rho_0^2 + (z_0 - z_2)^2} \tanh U_0 = f_{1,z_2}^0(\rho_0, z_0) \\ \tanh U_0 = \tanh U_0^a = \frac{k_0}{m} \\ V_0 = U_{2,z_1,z_2}(\rho_0, z_0), \quad \varepsilon_2 = 1 \\ C_3 = \Phi_{1,z_2}^0(\rho_0, z_0) \\ C_1 e^{2\gamma_0} = \frac{e^{2\gamma_K.}}{f_K.}\sqrt{\rho_0^2 + (z_0 - z_2)^2}, \end{cases} \qquad (3.99)$$

where $z_1 = -z_2 = k_0 = \sqrt{m^2 - a^2}$.

By a coordinate transformation (2.62) using (3.99), the Kerr metric in this region of spacetime can be approximately transformed into the Minkowski metric

$$ds_M^2 = d\rho'^2 + \rho'^2 d\varphi'^2 + dz'^2 - c^2 dt'^2.$$

With this approach, the Kerr metric in a small region of spacetime can be approximately regarded as a quasi-non-inertial reference frame (QNIRF).

2-I. Let us consider a small neighborhood of the region within this non-inertial reference frame:

$$\rho = \rho_0 = 0, \quad z = z_0$$

or

$$r = r_0, \quad \theta = 0, \qquad (3.100)$$

where (r, θ) are curvature coordinates.

It is easy to see that this case is similar to section (1.10), item (2.1), taking into account formulas (2.62)–(2.66).

2-II. The spiral motion in the quasi-inertial reference frame (QIRF) toward the gravitating center

$$\rho' = \rho'(\varphi'), \quad z' = 0, \quad \varphi' = \varphi'(t')$$

can be formally obtained from (2.62), (3.99) by setting

$$-z_1'(r_0) = e^{\gamma_0}\sqrt{2C_1\mu_+}\cosh\left[\frac{\sqrt{1+\tanh V_0}}{2}\frac{e^{-\gamma_0}}{C_1}(ct_{K.} - C_4\varphi_{K.})\right]$$

$$= \frac{2}{\sqrt{1+\tanh V_0}}\frac{C_1}{e^{-\gamma_0}}\sqrt{g_{00}}, \tag{3.101}$$

where

$$\mu_\pm = \sqrt{\rho^2 + (z-z_1)^2} \pm (z-z_1),$$

$$C_4 = C_3 + C_1\frac{\sqrt{1-\tanh V_0}}{\sqrt{1+\tanh V_0}}.$$

Thus, we obtain

$$\rho_0 \approx \rho(t), \quad z = z_0 = 0,$$

$$r_0 \approx r = r(t), \quad \theta = \frac{\pi}{2}. \tag{3.102}$$

In this case, we have

$$\rho' = \rho'(t'),$$

$$z' = 0,$$

$$t' = \frac{2\sqrt{g_{00}}}{\sqrt{1+\tanh V_0}}\frac{C_1}{e^{-\gamma_0}}\tanh\left[\frac{\sqrt{1+\tanh V_0}}{2}\frac{e^{-\gamma_0}}{C_1}(ct_{K.} - C_4\varphi_{K.})\right], \tag{3.103}$$

$$\varphi' = \varphi'(t') = \frac{e^{-\gamma_0}\varphi_{K.}}{\sqrt{1+\tanh V_0}} + \frac{\sqrt{1-\tanh V_0}}{2}\cdot\frac{e^{-\gamma_0}}{C_1}(ct_{K.} - C_4\varphi_{K.}).$$

At $t = 0$, it is easy to see that $dt' \approx dt\sqrt{g_{00}}$.

Considering that the Kerr metric 3.33 $(a_0 = -b_0)$ is approximately equal to the metric of the stationary Euclidon solution 2.58 in a small region of spacetime $\rho \approx \rho_0$, $z \approx z_0$, as is the Schwarzschild metric 1.35, 1.36 for the static Euclidon solution 1.83 by a coordinate transformation

$$\begin{cases} \dfrac{ct_{Sch.}e^{-\gamma_0'}}{C_1'\sqrt{2}} \approx \dfrac{\sqrt{1+\tanh V_0}}{2}\cdot\dfrac{e^{-\gamma_0}}{C_1}(ct_{K.} - C_4\varphi_{K.}) \\[3ex] \dfrac{e^{-\gamma_0}}{\sqrt{1+\tanh V_0}}\varphi_{K.} \approx \dfrac{e^{-\gamma_0'}}{\sqrt{2}}\varphi_{Sch.} - \dfrac{\sqrt{1-\tanh V_0}}{\sqrt{1+\tanh V_0}}\dfrac{ct_{Sch.}e^{-\gamma_0'}}{C_1'\sqrt{2}} \end{cases} \tag{3.104}$$

we can obtain (1.86) from (2.62), where $C_4 = C_3 + C_1\frac{\sqrt{1-\tanh V_0}}{\sqrt{1+\tanh V_0}}$,

In the case of a slight deviation from (3.104) spherical symmetry ($a \ll 1$), using ((1.89), (2.65)) we have

$$\varphi_{K.} \approx \varphi_{\mathrm{Sch.}} + \Omega_0' t_{\mathrm{Sch.}}, \tag{3.105}$$

where ((1.89), (2.65))

$$\Omega_0' = \frac{a_0'}{c} \sqrt{\frac{1 - \tanh V_0}{1 + \tanh V_0}}. \tag{3.106}$$

References

[1] J. Ehlers, *Collog. Theories Relative. Bruxelles* 49, (1959).

[2] J. Ozsvath, *Abhandl. Math. Natur. Cl. Acad. Wiss. ind Liter.* 1, 1 (1965).

[3] B.K. Harrison, *J. Math. Phys.* 9, 1744 (1968).

[4] R.J. Geroch, *J. Math. Phys.* 13, 393 (1972).

[5] W. Kinnersley, *J. Math. Phys.* 18, 1529 (1977).

[6] D. Maison, *Phys. Rev. Lett.* 41, (1978).

[7] W. Kinnersley, D.M. Chitre, *J. Math. Phys.* 18, 1538 (1977).

[8] W. Kirmersley, D.M. Chitre, *J. Math. Phys.* 19, 1926 (1978).

[9] W. Kinnersley, D.M. Chitre, *J. Math. Phys.* 19, 2037 (1978).

[10] W. Kinnersley, D.M. Chitre, *Phys. Rev. Lett.* 40, 1608 (1978).

[11] D.M. Chitre, *J. Math. Phys.* 19, 1626 (1978).

[12] D. Maison, *J. Math. Phys.* 20, 871 (1979).

[13] C. Hoenselaers, W. Kinnersley, B. Xanthopoulos, *J. Math. Phys.* 20, 2530 (1979).

[14] C. Hoenselaers, W. Kinnersley, B. Xanthopoulos, *Phys. Rev. Lett.* 42, 481 (1979).

[15] C. Hoenselaers, *J. Math. Phys.* 21, 2241 (1980).

[16] C. Hoenselaers, *J. Phys. A: Math. Gen.* 14, 427 (1981).

[17] W. Dietz, C. Hoenselaers, *Proc. R. Soc. Lond.* A382, 221 (1982).

[18] W. Dietz, *Gen. Rel. Grav.* 15, 911 (1983).

[19] C. Hoenselaers, W. Dietz, *Gen. Rel. Grav.* 16, 71 (1984).

[20] W. Dietz, *Gen. Rel. Grav.* 16, 249 ((1984).

[21] I. Hauser, F.J. Ernst, *Phys. Rev.* D20, 365 (1979).

[22] I. Hauser, F.J. Ernst, *Phys. Rev.* D20, 1783 (1979).

[23] I. Hauser, F.J. Ernst, *J. Math. Phys.* 21, 1126 (1980).

[24] I. Hauser, F.J. Ernst, *J. Math. Phys.* 21, 1418 (1980).

[25] I. Hauser, F.J. Ernst, *J. Math. Phys.* 22, 1051 (1981).

[26] C. Hoenselaers, Axisyminetric stationary vacuum solutions of Einstein's equations, Habilitationsschrift, Universitat Munchen (1982).

[27] C. Hoenselaers, *J. Phys. A: Math. Gen.* 15, 3531 (1982).

[28] W. Dietz, C. Hoenselaers, *Phys. Rev. Lett.* 48, 778 (1982).

[29] See: R. Ruffini (ed.), *Proceedings of 2nd Marcel Grossman Meeting of General Relativity*, North-Holland, Amsterdam (1982).

[30] K. Veno, *Publ. RIMS. Kyoto University* 19, 59 (1983).

[31] Y. Wu, M. Ge, *J. Math. Phys.* 24, 1187 (1983).

[32] See: E.R. Ruffini (ed.), *Proceedings of 3rd Marcel Grossman Meeting of General Relativity*, North-Holland, Amsterdam (1983).

[33] W. Dietz, On techniques for generating solutions and generated two mass solutions of Einstein's vacuum equations, Habilitationsschrift, Universitat Wurzburg (1984).

[34] C. Hoenselaers, N. Dadhich, *Class. Quant. Grav.* 1, 389 (1984).

[35] H. Quevedo, B. Mashhoon, *Phys. Lett.* A109, 13 (1985).

[36] See: R. Ruffini (ed.), *Proceedings of 4th Marsel Grossman Meeting of General Relativity*, North-Holland, Amsterdam (1985).

[37] H. Quevedo, *Phys. Rev.* D33, 324 (1986).

[38] H. Quevedo, Ph. D. Thesis, University of Cologne (1987).

[39] H. Quevedo, *Phys. Rev.* D39, 2904 (1989).

[40] H. Quevedo, *Fortschr. Phys.* 38, 733 (1990).

[41] H. Quevedo, B. Mashhoon, *Phys. Lett.* A148, 149 (1990).

[42] H.Quevedo *Phys. Rev. Lett.* 67, 1050 (1991).

[43] H. Quevedo, *Phys. Rev.* D43, 3902 (1993).

[44] A. Belinsky, V.-E. Zakharov, *Sov. Phys. JETP* 48, 985 (1978),

[45] V.A. Belinsky, V.E. Zakharov, *Sov. Phys. JETP* 50, 1 (1979).

[46] R.T. Jantzen, *Nuovo Cim.* B59, 287 (1980).

[47] G.A. Alekseev, *Sov. Phys. Dokl.* 26, 158 (1981).

[48] G. Neugebauer, *Phys. Lett.* A86, 91 (1981).

[49] V.A. Belinsky, M. Prancaviglia, *Gen. Rel. Grav.* 14, 213 (1982),

[50] E. Verdaguer, *J. Phys. A: Math. Gen.* 15,1261 (1982).

[51] M. Yamazaki, *Phys. Lett.* 50, 1027 (1983).

[52] B.J. Carr, E.B. Verdaguer, *Phys. Rev.* D28, 2995 (1983).

[53] J. Ibanez, E. Verdaguer, *Phys. Rev. Lett.* 51, 1313 (1983).

[54] A.P. Veselov, *Theor. Math. Phys.* 54, 155 (1983).

[55] R.J. Gleiser, *Gen. Rel. Grav.* 16, 1077 (1984).

[56] P.S. Letelier, *J. Math. Phys.* 26, 467 (1985).

[57] J. Ibanez, E. Verdaguer, *Phys. Rev.* D31, 251 (1985).

[58] K.C. Das, *Phys. Rev.* D31, 927 (1985).

[59] G.A. Alekseev, *Gravit. Electromagn.* 4, 3 (1989).

[60] A. Dagotto, R. Gleiser, G. Conzalez, J. Pullin, *Phys. Lett.* A146, 15 (1990).

[61] S.L. Tertychniy, *Class. Quant. Grav.* 7, 1345 (1990).

[62] P.T. Boyd, J.M. Centrella, S.A. Klasky, *Phys. Rev.* D43, 79 (1991).

[63] V.A. Belinsky, *Phys. Rev.* D44, 3109 (1991).

[64] Y. Wang, Z. He, *J. Shanghai Jiaotong Univ.* 28, 62 (1994).
[65] A. Nakamura, *J. Phys. Soc. Jpn.* 63, 1214 (1994).
[66] B.K. Harrison, *Phys. Rev. Lett.* 41, 1197 (1978).
[67] G. Neugebauer, *J. Phys. A: Math. Gen.* 12, 67 (1979).
[68] G. Neugebauer, D. Kramer, *Exp. Tech. Phys.* 28, 3 (1980).
[69] G. Neugebauer, *J. Phys. A: Math, Gen.* 13, 19 (1980).
[70] G. Neugebauer, *J. Phys, A: Math. Gen.* 13, 1737 (1980).
[71] D. Kramer, G. Neugebauer, *Phys. Lett.* A75, 259 (1980).
[72] B.K. Harrison, *Phys. Rev.* D21, 1695 (1980).
[73] E. Kyriakopoulos, *Gen. Relat. Grav.* 12, 989 (1980).
[74] M. Omote, M. Wadati, *J. Math. Phys.* 22, 961 (1981).
[75] F.J. Chinea, *Phys. Rev.* D24, 1053(1981).
[76] F. Fayos, E. Fornells, J.A. Lobo, *Gen. Rel. Grav.* 11, 1093 (1989).
[77] C.M. Cosgrove, *J. Math. Phys.* 21, 2417 (1980).
[78] C.M. Cosgrove, *J. Math. Phys.* 22, 2624 (1981).
[79] C.M. Cosgrove, *J. Math. Phys.* 23, 615 (1982).
[80] F.J. Chinea, *Phys. Rev.* D26, 2175 (1982).
[81] M. Omote, M. Wadati, *Physica* A114, 100 (1982).
[82] F.J. Chinea, *Physica* A114, 151 (1982).
[83] F.J. Chinea, *Phys. Rev. Lett.* 50, 221 (1983).
[84] B.K. Harrison, *J. Math. Phys.* 24, 2178 (1983).
[85] M. Gurses, *Phys. Rev. Lett.* 51, 1810 (1983).
[86] R.S. Ward, *Gen. Rel. Grav.* 15, 105 (1983).
[87] R.S. Ward, *Phys. Lett.* A100, 349 (1984).
[88] M. Gurses, *Phys. Lett.* A101, 388 (1984).
[89] E. Kyriakopoulos, *J. Phys. A: Math. Gen.* 20, 1667 (1987).
[90] E. Kyriakopoulos, *Gen. Relat. Grav.* 20, 1067 (1988).
[91] E. Kyriakopoulos, *Class. Quant. Grav.* 6, 35 (1989).
[92] E. Kyriakopoulos, *Int. J. Mod. Phys.* A5, 1285 (1990).
[93] Y. Wu, *Int. J. Theor. Phys.* 31, 161 (1992).
[94] M. Gurses, *Phys. Rev. Lett.* 70, 367 (1993).
[95] A. Tomimatsu, H. Sato, *Supp. Prog. Theor. Phys.* 70, 215 (1981).
[96] K. Oohara, H. Sato, *Prog. Theor. Phys.* 65, 1891 (1981).
[97] A. Tomimatsu, M. Kihara, *Prog. Theor. Phys.* 67, 1406 (1982).
[98] M. Yamazaki, *Prog. Theor. Phys.* 69, 503 (1983).
[99] A. Tomimatsu, *Prog. Theor. Phys.* 70, 385 (1983).
[100] C. Hoenselaers, *Prog. Theor. Phys.* 72, 761 (1984).
[101] A.T. Ilitchev, N.R. Sibgatullin, *Izv. Vuzov. Ser. Fizika* 9, 75 (1986)
[102] D.A. Korotkin, *Theor. Matem. Fizika* 77, 25 (1988).
[103] T.I. Gutsunaev, V.S. Manko, S.M. Abramyan, *Problemi Statist. Fizitg Teor. Polya* (in Russian) 129 (1987).
[104] T.I. Gutsunaev, V.S. Manko, *Gen. Rel. Grav.* 20, 327 (1988),

[105] S.B. Beisekeev, T.I. Gutsunaev, *Pisma Zh. Exsp. Theor. Fiz.* 54, 597

[106] T.I. Gutsunaev, V.B. Borsov, S.L. Elsgolts, *Questions of electro-dynamics in spaces with a non-Galilean metric*, TSTU-Press, Tver (1980) (in Russian).

[107] T.I. Gutsunaev, V.A Chernyaev, *Axisymmetric gravitational fields*, MXCA-Press, Moscow (2004).

[108] T.I. Gutsunaev, V.A. Chernyaev, S.L. Elsgolts, *Gravit. Cosmol.* 5, 335 (1999).

[109] T.I. Gutsunaev, A.A. Shaideman, S.L. Elsgolts, *Gravit. Cosmol.* 6, 254 (2000).

[110] J. Gariel, G. Marcilhacy, *Gravit. Cosmol.* 8, 209 (2002).

[111] A.A. Shaideman, Generation of vacuum axially symmetric solutions to the equations of General Relativity using stationary Euclidons, PhD thesis, Moscow (2005) (in Russian).

[112] A.A. Shaideman, K.V. Golubnichiy, Stationary multiple euclidon solutions to the vacuum Einstein equations, *arXiv preprint* (2025). arXiv:2502.03675, *Commun. Anal. Comput. (CAC)* (submitted 2024).

Chapter 4

Static Solutions of the Einstein–Maxwell Equations

4.1 Introduction

4.2 Basic Equations

The metric of the static axisymmetric gravitational field can be written in the canonical Weyl coordinates in the form

$$ds^2 = f^{-1}[e^{2\gamma}(d\rho^2 + dz^2) + \rho^2 d\varphi^2] - f dt^2, \qquad (4.1)$$

where the functions f and γ depend only on ρ and z.

The Einstein–Maxwell equations have the form

$$R_{ik} = 8\pi T_{ik},$$

$$\frac{\partial}{\partial x^k}(\sqrt{-g}F_{lm}g^{il}g^{km}) = 0,$$

$$\frac{\partial F_{ik}}{\partial x^l} + \frac{\partial F_{kl}}{\partial x^i} + \frac{\partial F_{li}}{\partial x^k} = 0,$$

where R_{ik} is the Einstein–Ricci tensor and T_{ik} is the energy–momentum tensor of the electromagnetic field, connected with the electromagnetic field intensity tensor F_{ik} by the relation

$$T_{ik} = \frac{1}{4\pi}\left(F_{il}F_k^l - \frac{1}{4}F_{lm}F^{lm}g_{ik}\right).$$

It is also customary to introduce the 4-potential of the electromagnetic field

$$F_{ik} = \frac{\partial A_k}{\partial x^i} - \frac{\partial A_i}{\partial x^k}.$$

113

The fact that the static Einstein–Maxwell equations allow the existence of either the electric potential, or the magnetic one, results from the stationary Einstein–Maxwell equations.

In this connection, we set $A_i = [0, 0, A_3(\rho, z), 0]$ and rewrite equations explicitly as

$$f\Delta f = (\vec{\nabla} f)^2 + 2f^3\rho^{-2}(\vec{\nabla} A_3)^2, \quad \vec{\nabla}\left(f\rho^{-2}\vec{\nabla} A_3\right) = 0, \quad (4.2)$$

$$4\frac{\partial\gamma}{\partial\rho} = \frac{\rho}{f^2}\left[\left(\frac{\partial f}{\partial\rho}\right)^2 - \left(\frac{\partial f}{\partial z}\right)^2\right] + \frac{4f}{\rho}\left[\left(\frac{\partial A_3}{\partial\rho}\right)^2 - \left(\frac{\partial A_3}{\partial z}\right)^2\right],$$
$$(4.3)$$

$$2\frac{\partial\gamma}{\partial z} = \frac{\rho}{f^2}\frac{\partial f}{\partial\rho}\frac{\partial f}{\partial z} + \frac{4f}{\rho}\frac{\partial A_3}{\partial\rho}\frac{\partial A_3}{\partial z},$$

where

$$\Delta \equiv \frac{\partial^2}{\partial\rho^2} + \frac{1}{\rho}\frac{\partial}{\partial\rho} + \frac{\partial^2}{\partial z^2}, \quad \vec{\nabla} \equiv \vec{\rho_0}\frac{\partial}{\partial\rho} + \vec{z_0}\frac{\partial}{\partial z},$$

where $\vec{\rho_0}$ and $\vec{z_0}$ are unit vectors and $A_3(\rho, z)$ is the magnetic component of the electromagnetic 4-potential.

The second equation in (4.2) can be viewed as a condition for the existence of a new potential A'_3, connected with A_3 by the relations

$$\frac{\partial A'_3}{\partial\rho} = -\frac{f}{\rho}\frac{\partial A_3}{\partial z}, \quad \frac{\partial A'_3}{\partial z} = \frac{f}{\rho}\frac{\partial A_3}{\partial\rho}. \quad (4.4)$$

In that case, equations (4.2) and (4.3) can be rewritten as

$$f\Delta f = (\vec{\nabla} f)^2 + 2f(\vec{\nabla} A'_3)^2, \quad f\Delta A'_3 = \vec{\nabla} A'_3\vec{\nabla} f, \quad (4.5)$$

$$4\frac{\partial\gamma}{\partial\rho} = \frac{1}{f^2}\left[\left(\frac{\partial f}{\partial\rho}\right)^2 - \left(\frac{\partial f}{\partial z}\right)^2\right] + \frac{4}{f}\left[\left(\frac{\partial A'_3}{\partial z}\right)^2 - \left(\frac{\partial A'_3}{\partial\rho}\right)^2\right],$$
$$(4.6)$$

$$\frac{2}{\rho}\frac{\partial\gamma}{\partial z} = \frac{1}{f^2}\frac{\partial f}{\partial\rho}\frac{\partial f}{\partial z} - \frac{4}{f}\frac{\partial A'_3}{\partial\rho}\frac{\partial A'_3}{\partial z}.$$

On can easily see that the electrostatic Einstein–Maxwell equations have the same form as equations (4.5). Therefore, we can put

$A_4 = A'_3$, where A_4 is the electric component of the electromagnetic 4-potential $A_i = [0, 0, 0, -A_4(\rho, z)]$.

Hence, one can obtain and electrostatic analogue of known solutions to equations (4.2)–(4.4). And reversely, one can put into correspondence to any solution of equations (4.5) an appropriate magnetostatic Einstein–Maxwell solution.

If we introduce further the function $u = \sqrt{f}$, we obtain the equations

$$u\Delta u = (\vec{\nabla} u)^2 + (\vec{\nabla} A'_3)^2, \quad u\Delta A'_3 = 2\vec{\nabla} u \vec{\nabla} A'_3. \tag{4.7}$$

In the magnetic case, these equations have the forms

$$u\Delta u = (\vec{\nabla} u)^2 + \frac{u^4}{\rho^2}(\vec{\nabla} A_3)^2, \quad \vec{\nabla}\left(\frac{u^2}{\rho^2}\vec{\nabla} A_3\right)^2 = 0. \tag{4.8}$$

The magnetostatic equations (4.8) can be rewritten in the form

$$F\Delta F = (\vec{\nabla} F)^2 - (\vec{\nabla} A_3)^2, \quad \vec{\nabla}\left(\frac{\vec{\nabla} A_3}{F^2}\right) = 0 \tag{4.9}$$

by substitution of $u = \frac{\rho}{F}$.

From equations (4.7), it is not difficult to go over to an equivalent system of differential equations by introducing the potentials ε_1 and ε_2, by the formulae

$$\epsilon_1 = u + A'_3, \quad \epsilon_2 = u - A'_3, \tag{4.10}$$

yielding as a result a formulation of the electrostatic (magnetostatic) problem in a symmetric form with respect to ε_1 and ε_2

$$(\epsilon_1 + \epsilon_2)\Delta\epsilon_1 = 2(\vec{\nabla}\epsilon_1)^2, \quad (\epsilon_1 + \epsilon_2)\Delta\epsilon_2 = 2(\vec{\nabla}\epsilon_2)^2. \tag{4.11}$$

We note certain general properties of equations (4.11). Let $(\tilde{\epsilon}_1, \tilde{\epsilon}_2)$ denote some solution of equations (4.11). Then, the functions

$$\epsilon_1 = \frac{-a_0 + b_0\tilde{\epsilon}_1}{c_0 - d_0\tilde{\epsilon}_1}, \quad \epsilon_2 = \frac{a_0 + b_0\tilde{\epsilon}_2}{c_0 + d_0\tilde{\epsilon}_2}, \tag{4.12}$$

where a_0, b_0, c_0 and d_0 are arbitrary real constants, will also satisfy equations (4.11).

In the particular case when we set in (4.12) $c = 0, a = -2dc_0$ and $b = -d$, we obtain

$$\epsilon_1 = -\frac{2c_0}{\tilde{\epsilon}_1} + 1, \quad \epsilon_2 = -\frac{2c_0}{\tilde{\epsilon}_2} - 1. \tag{4.13}$$

The analog of transformations (4.13) for the vacuum stationary gravitational field is given by the inversion transformations (see, e.g., Ref. [29]).

The resultant abbreviated transformation is the Ehlers transformation.

In this way, given a certain solution $(\tilde{f}, \tilde{A}'_3)$ of equations (4.5), a new solution can be constructed as

$$f = 4c_0^2 \tilde{f} \left(\tilde{A}_3'^2 - \tilde{f} \right)^{-2}, \quad A'_3 = 1 - 2c_0 \tilde{A}'_3 \left(\tilde{A}_3'^2 - \tilde{f} \right)^{-1}. \tag{4.14}$$

Further, it follows from (4.6) and (4.13) that $\gamma = \tilde{\gamma}$.

In the functions u and A'_3, expression (4.12) take the form

$$u = \frac{(b_0 c_0 - a_0 d_0)\tilde{u}}{(c_0 - d_0 \tilde{A}'_3)^2 - d_0^2 \tilde{u}^2},$$

$$A'_3 = \frac{-a_0 c_0 + (a_0 d_0 + b_0 c_0)\tilde{A}'_3 + b_0 d_0(\tilde{u}^2 - \tilde{A}_3'^2)}{(c_0 - d_0 \tilde{A}'_3)^2 - d_0^2 \tilde{u}^2}. \tag{4.15}$$

Obviously, the similar theorem takes place for equations (4.9).

Let $\tilde{F}$ and $\tilde{A}_3$ be an arbitrary solution of equation (4.9). Then, the functions

$$F = \frac{(\beta_0 \gamma_0 + \alpha_0 \delta_0)\tilde{F}}{(\gamma_0 - \delta_0 \tilde{A}_3)^2 + \delta_0^2 \tilde{F}^2},$$

$$A_3 = \frac{-\alpha_0 \gamma_0 - (\beta_0 \gamma_0 - \alpha_0 \delta_0)\tilde{A}_3 + \beta_0 \delta_0(\tilde{F}^2 + \tilde{A}_3^2)}{(\gamma_0 - \delta_0 \tilde{A}_3)^2 + \delta_0^2 \tilde{F}^2}, \tag{4.16}$$

where $\alpha_0, \beta_0, \gamma_0$ and δ_0 are arbitrary real constants. By setting $\alpha_0 = 0, \beta_0 = 1, \gamma_0 = 1, \delta_0 = c_0$, (4.16) yields

$$F = \frac{\tilde{F}}{(1 - c_0 \tilde{A}_3)^2 + c_0^2 \tilde{F}^2}, \quad A_3 = \frac{\tilde{A}_3 - c_0(\tilde{F}^2 + \tilde{A}_3^2)}{(1 - c_0 \tilde{A}_3)^2 + c_0^2 \tilde{F}^2}. \tag{4.17}$$

In conclusion of this section, we give the formulation of an important Bonnor's theorem [28] establishing the relationship between stationary vacuum and static electrovacuum axisymmetric solutions. The essential meaning of this theorem is revealed when one considers stationary vacuum equations

$$f\Delta f = (\vec{\nabla} f)^2 - (\vec{\nabla}\Phi)^2, \quad f\Delta\Phi = 2\vec{\nabla} f\,\vec{\nabla}\Phi \qquad (4.18)$$

and equations (4.7) for the electrostatic (magnetostatic) field. With the transformation $A'_3 \to iA'_3$, both systems become identical. Hence, if the vacuum stationary solution is found, i.e., if are determined the real functions f and Φ satisfying equations (4.18), then the corresponding electrovacuum metric function f_{el} is equal to f^2, and the potential A_4, (or A'_3) is equal to $i\Phi$. The obtained static electrovacuum solution will be real only if one can carry out in it a complex continuation of the parameters.

4.3 The Weyl–Bonnor–Papapetrou–Majumdar Class of Solutions

This class is characterized by the functional connection in (4.7) to each other of the potentials u and A'_3 so that $u = u(A'_3)$.

Let us rewrite the system of equations (4.8) in the form

$$\vec{\nabla}\left[\frac{\vec{\nabla}(u^2 - A'^2_3)}{u^2}\right] = 0, \quad \vec{\nabla}\left(\frac{\vec{\nabla} A'_3}{u^2}\right) = 0 \qquad (4.19)$$

and introduce an arbitrary harmonic function ψ so that $u = u(\psi)$ and $A'_3 = A'_3(\psi)$. In such a case, equations (4.19) are simplified and can be integrated as

$$\frac{d}{d\psi}(u^2 - A'^2_3) = c_1 u^2, \quad \frac{dA'_3}{d\psi} = c_2 u^2,$$

$$\psi = \int \frac{dA'_3}{c_2 A'^2_3 + c_1 A'_3 + c_3}. \qquad (4.20)$$

Different relations between constants c_1, c_2 and c_3 give different subclasses of solution (4.20).

1. $c_1^2 > 4c_2c_3$ gives the subclass of the Weyl electrovacuum solutions

$$u = \frac{(1 + a_0^2)e^\psi}{1 - a_0^2 e^{2\psi}}, \quad A_3' = \frac{a_0(1 + e^{2\psi})}{1 - a_0^2 e^{2\psi}}. \tag{4.21}$$

While $a_0 = 0$, the solution (4.21) turns to the Weyl vacuum static gravitational field.

The solution (4.21) includes also the most famous Weyl–Reissner–Nordstrom spherical-symmetric solution. It is easy to obtain it by choosing the function

$$\psi = \frac{1}{2} \ln \frac{x - 1}{x + 1} \quad \text{and setting} \quad a_0 = \frac{\sqrt{m_0 - k_0}}{\sqrt{m_0 + k_0}}, \quad k_0^2 = m_0^2 - e^2.$$

Then,

$$f = u^2 = \frac{m_0^2(x^2 - 1)}{(k_0 x + m_0)^2}, \quad A_4 = \frac{e}{k_0 x + m_0}. \tag{4.22}$$

By the mapping $r = k_0 x + m_0$, we have

$$f = 1 - \frac{2m_0}{r} + \frac{e^2}{r^2}, \quad A_4 = \frac{e}{r}. \tag{4.23}$$

It seems that solution (4.23) describes the external gravitational field of the charged mass.

2. $c_1^2 < 4c_2c_3$ gives the Bonnor subclass of the electrovacuum solutions [7]:

$$u = \frac{1}{\cos \psi}, \quad A_3' = \tan \psi. \tag{4.24}$$

It was used to describe the gravitational field of a massless dipole. By choosing $\psi = \mu_0 \frac{z}{(\rho^2 + z^2)^{3/2}}$, it is easy to obtain

$$f = u^2 = \cos^{-2} \frac{\mu_0 z}{(\rho^2 + z^2)^{3/2}}, \quad A_3' = \tan \frac{\mu_0 z}{(\rho^2 + z^2)^{3/2}}. \tag{4.25}$$

In the coordinates $\rho = R \sin \theta$, $z = R \cos \theta$, it has asymptotics

$$f \to 1 + \frac{\mu_0^2 \cos^2 \theta}{R^4}, \quad A_3' \to \frac{\mu_0 \cos \theta}{R^2}.$$

This solution was used by Kramer [30]. The function ψ was chosen of the other form. In the (r, θ) coordinates, the solution becomes

$$f = \cos^{-2}\left(\frac{\mu_0 \cos\theta}{r^2 - 2\mu_0 r + \mu_0^2 \sin^2\theta}\right),$$

$$A_3 = \frac{\mu_0(r - \mu_0)\sin^2\theta}{r^2 - 2\mu_0 r + \mu_0^2 \sin^2\theta}. \tag{4.26}$$

Note also that this subclass was used as nonlinear superposition with Kerr solution in order to describe the gravitational field of a mass endowed with magnetic dipole moment.

3. $c_1^2 = 4c_2 c_3$ gives the subclass of Papapetrou–Majumdar solutions (4.5):

$$u = \frac{1}{1 + \psi}, \quad A_3' = \frac{\psi}{1 + \psi}. \tag{4.27}$$

Tauber [9] supposed it to describe the gravitational field of a massless magnetic dipole.

For $\psi = \mu_0 \frac{z}{(\rho^2 + z^2)^{3/2}}$, the gravitational and magnetic potentials have the forms

$$f = \left[1 + \frac{\mu_0 z}{(\rho^2 + z^2)^{3/2}}\right]^{-2},$$

$$A_3' = \frac{\mu_0 z}{(\rho^2 + z^2)^{3/2}}\left[1 + \frac{\mu_0 z}{(\rho^2 + z^2)^{3/2}}\right]^{-1}. \tag{4.28}$$

According to this formula, the gravitational potential depends on the sign of μ_0.

4.4   The Herlt Class of Solutions

Let us take Lewis solution of the equations (4.9)

$$F^L = \frac{\rho}{\alpha_0}\sinh\left(\frac{\alpha_0}{\rho}\frac{\partial\chi}{\partial\rho}\right), \quad A_3^L = \frac{\partial\chi}{\partial z}, \tag{4.29}$$

where $\alpha_0 = \text{const}$ and χ satisfies the equation

$$\Delta'\chi \equiv \frac{\partial^2\chi}{\partial\rho^2} - \frac{1}{\rho}\frac{\partial\chi}{\partial\rho} + \frac{\partial^2\chi}{\partial z^2} = 0. \tag{4.30}$$

Using transformations (4.16), we can construct a new class of solutions:

$$F = \frac{F^L}{(F^L)^2 + (A_3^L)^2}, \quad A_3 = \frac{A_3^L}{(F^L)^2 + (A_3^L)^2}, \quad u = \frac{\rho}{F}. \qquad (4.31)$$

For this solution, the potential A_3' can be found by integration of equations (4.4), yielding

$$A_3' = 2\chi - \rho\frac{\partial\chi}{\partial\rho} - \left(\frac{\partial\chi}{\partial z}\right)^2 \alpha_0 \coth\left(\frac{\alpha_0}{\rho}\frac{\partial\chi}{\partial\rho}\right). \qquad (4.32)$$

Therefore, our solution which satisfies equations (4.7) is given by the following relations:

$$\tilde{u} = \frac{\rho^2}{\alpha_0}\sinh\left(\frac{\alpha_0}{\rho}\frac{\partial\chi}{\partial\rho}\right) + \left(\frac{\partial\chi}{\partial z}\right)^2 \alpha_0 \sinh^{-1}\left(\frac{\alpha_0}{\rho}\frac{\partial\chi}{\partial\rho}\right),$$

$$\tilde{A}_3' = 2\chi - \rho\frac{\partial\chi}{\partial\rho} - \left(\frac{\partial\chi}{\partial z}\right)^2 \alpha_0 \coth\left(\frac{\alpha_0}{\rho}\frac{\partial\chi}{\partial\rho}\right). \qquad (4.33)$$

Then, according to (4.15), one is able to construct the solution

$$u = \frac{c_1 \tilde{u}}{\tilde{u}^2 - \tilde{A}_3'^2}, \quad A_3' = \frac{c_1 \tilde{A}_3'}{\tilde{u}^2 - \tilde{A}_3'^2} + c_2, \qquad (4.34)$$

where c_1 and c_2 are arbitrary real constants. For $c_1 = -4$, $c_2 = -1$, the solution (4.34) of equations (4.7) in accordance with (4.33) can be finally written in the form

$$u + A_3' = 4\left[2\chi - \rho\frac{\partial\chi}{\partial\rho} - \frac{\rho^2}{\alpha_0}\sinh\left(\frac{\alpha_0}{\rho}\frac{\partial\chi}{\partial\rho}\right)\right.$$
$$\left. - \left(\frac{\partial\chi}{\partial z}\right)^2 \cdot \alpha_0 \coth\left(\frac{\alpha_0}{2\rho}\frac{\partial\chi}{\partial\rho}\right)\right]^{-1} - 1,$$

$$u - A_3' = -4\left[2\chi - \rho\frac{\partial\chi}{\partial\rho} + \frac{\rho^2}{\alpha_0}\sinh\left(\frac{\alpha_0}{\rho}\frac{\partial\chi}{\partial\rho}\right)\right.$$
$$\left. - \left(\frac{\partial\chi}{\partial z}\right)^2 \cdot \alpha_0 \tanh\left(\frac{\alpha_0}{2\rho}\frac{\partial\chi}{\partial\rho}\right)\right]^{-1} + 1. \qquad (4.35)$$

It is the Herlt class of solutions [18]. While the parameter $a_0 \to 0$, one can obtain the "cut-solution" Herlt class [17]:

$$u + A_3' = 2\left[\chi - \frac{(\vec{\nabla}\chi)^2}{\frac{1}{\rho}\frac{\partial\chi}{\partial\rho}}\right]^{-1} - 1, \quad u - A_3' = -\frac{2}{\chi} + 1. \tag{4.36}$$

In the prolate ellipsoidal coordinates (x, y), we have

$$(\vec{\nabla}\chi)^2 = \frac{1}{k_0(x^2 - y^2)}\left[(x^2 - 1)\left(\frac{\partial\chi}{\partial x}\right)^2 + (1 - y^2)\left(\frac{\partial\chi}{\partial y}\right)^2\right],$$

$$\frac{1}{\rho}\frac{\partial\chi}{\partial\rho} = \frac{1}{k_0(x^2 - y^2)}\left[x\frac{\partial\chi}{\partial x} - y\frac{\partial\chi}{\partial y}\right],$$

$$(x^2 - 1)\frac{\partial^2\chi}{\partial x^2} + (1 - y^2)\frac{\partial^2\chi}{\partial y^2} = 0,$$

and choosing the function χ in the form $\chi = px + qy + 1$, when $p^2 - q^2 = 1$, we find from (4.36)

$$f = u^2 = \left[\frac{p^2x^2 - q^2y^2 - 1}{(px + 1)^2 - q^2y^2}\right]^2,$$

$$A_3' = \frac{-2qy}{(px + 1)^2 - q^2y^2}. \tag{4.37}$$

Formulae (4.37) define the Bonnor [10] solution for a massive magnetic (or electric) dipole.

4.5 The Special Class of Solutions

We rewrite equations (4.5) in the form

$$f\left[\frac{1}{\rho}\frac{\partial}{\partial\rho}\left(\rho\frac{\partial f}{\partial\rho}\right) + \frac{\partial^2 f}{\partial z^2}\right]$$

$$= \left(\frac{\partial f}{\partial\rho}\right)^2 + \left(\frac{\partial f}{\partial z}\right)^2 + 2f\left[\left(\frac{\partial A_3'}{\partial\rho}\right)^2 + \left(\frac{\partial A_3'}{\partial z}\right)^2\right],$$

$$f\left[\frac{1}{\rho}\frac{\partial}{\partial\rho}\left(\rho\frac{\partial A_3'}{\partial\rho}\right) + \frac{\partial^2 A_3'}{\partial z^2}\right] = \frac{\partial f}{\partial\rho}\frac{\partial A_3'}{\partial\rho} + \frac{\partial f}{\partial z}\frac{\partial A_3'}{\partial z} \tag{4.38}$$

and introduce into consideration the functions ψ and χ satisfying the equations

$$\frac{\partial^2 \psi}{\partial \rho^2} + \frac{1}{\rho}\frac{\partial \psi}{\partial \rho} + \frac{\partial^2 \psi}{\partial z^2} = 0, \quad \frac{\partial^2 \chi}{\partial \rho^2} - \frac{1}{\rho}\frac{\partial \chi}{\partial \rho} + \frac{\partial^2 \chi}{\partial z^2} = 0 \qquad (4.39)$$

and also connected to each other by the relations

$$\frac{\partial \psi}{\partial \rho} = \frac{1}{\rho}\frac{\partial \chi}{\partial z}, \quad \frac{\partial \psi}{\partial z} = -\frac{1}{\rho}\frac{\partial \chi}{\partial \rho}. \qquad (4.40)$$

It is not hard to show that we can take as the solutions $\left(\tilde{A}_3, \tilde{f}\right)$ of equations (4.38) the following functions:

$$\tilde{A}'_3 = -\chi, \quad \tilde{f} = \rho^2 \cosh^2 \psi. \qquad (4.41)$$

In accordance with (4.15), we can construct a large class of solutions [25]:

$$f = \frac{(b_0 c_0 - a_0 d_0)^2 \rho^2 \cosh^2 \psi}{[c_0^2 + 2c_0 d_0 \chi - d_0^2(\rho^2 \cosh^2 \psi - \chi^2)]^2},$$

$$A_3 = \frac{-a_0 c_0 + b_0 d_0(\rho^2 \cosh^2 \psi - \chi^2) - (a_0 d_0 + b_0 c_0)\chi}{c_0^2 + 2c_0 d_0 \chi - d_0^2(\rho^2 \cosh^2 \psi - \chi^2)}, \qquad (4.42)$$

where a_0, b_0, c_0 and d_0 are real constants satisfying the condition $b_0 c_0 - a_0 d_0 \neq 0$.

We consider the most interesting cases of this class of solutions:

1. $a_0 = d_0 = 0$, $b_0 = c_0 \mu_0$, $\psi = \ln\left(\frac{\mu_0 \rho}{2}\right)$, $\chi = z$.

In this case, from (4.42) with (4.39), (4.40) and (4.4), it follows

$$f = \left(1 + \frac{\mu_0^2}{4}\rho^2\right)^2,$$

$$A'_3 = A_4 = -\mu_0 z, \quad A_3 = \frac{\mu_0}{2}\rho^2\left(1 + \frac{\mu_0^2}{4}\rho^2\right)^{-1}, \qquad (4.43)$$

i.e., we obtain the Bonnor–Melvin solution [7, 8] describing a homogeneous magnetic (electric) universe.

We consider as well the equations of the lines of force of the magnetic and electric fields [31]. For the magnetic field, in particular, we obtain

$$\frac{\sqrt{g_{11}}dx^{(1)}}{H_1} = \frac{\sqrt{g_{22}}dx^{(2)}}{H_2} = \frac{\sqrt{g_{33}}dx^{(3)}}{H_3}, \qquad (4.44)$$

where the Cartesian components of the magnetic field intensity can be written as

$$H_1 = \frac{F_{23}}{\sqrt{g_{22}}\sqrt{g_{33}}}, \qquad H_2 = -\frac{F_{13}}{\sqrt{g_{11}}\sqrt{g_{33}}}, \qquad H_3 = \frac{F_{12}}{\sqrt{g_{11}}\sqrt{g_{22}}}, \qquad (4.45)$$

where $F_{ik} = \frac{\partial A_k}{\partial x^i} - \frac{\partial A_i}{\partial x^k}$ is the electro-magnetic field tensor.

The equations of the lines force for the electric field can be obtained from (4.44) by the formal replacement $H_1 \to E_1$, $H_2 \to E_2$, $H_3 \to E_3$ but the Cartesian components of the electric field are written as

$$E_1 = \frac{F_{10}}{\sqrt{g_{11}}\sqrt{-g_{00}}}, \qquad E_2 = \frac{F_{20}}{\sqrt{g_{22}}\sqrt{-g_{00}}}, \qquad E_3 = \frac{F_{30}}{\sqrt{g_{33}}\sqrt{-g_{00}}}. \qquad (4.46)$$

The equations of the lines of force in the coordinates (R, θ) ($\rho = R\sin\theta$, $z = R\cos\theta$) with (4.44) and (4.45) take the form

$$\frac{dR}{d\theta} = -R\cot\theta, \qquad \frac{d\theta}{d\psi} = 0. \qquad (4.47)$$

And exact solution of equations (4.47) is $R\sin\theta = C_0$, where C_0 is the constant characterizing the family of the lines of force.

2. $c_0 = 0$, $b_0 = d_0$, $a_0 = 2(m + e)d_0$, $k_0^2 = m^2 - e^2$,

$$\chi = k_0 z + m + e, \qquad \psi = \frac{1}{2}\ln[(1 + y)/(1 - y)].$$

Here, the functions ψ and χ satisfy the equations

$$\frac{\partial}{\partial x}\left[(x^2 - 1)\frac{\partial\psi}{\partial x}\right] + \frac{\partial}{\partial y}\left[(1 - y^2)\frac{\partial\psi}{\partial y}\right] = 0,$$

$$(x^2 - 1)\frac{\partial^2\chi}{\partial x^2} + (1 - y^2)\frac{\partial^2\chi}{\partial y^2} = 0,$$

$$\frac{\partial\psi}{\partial x} = -\frac{1}{k_0(x^2 - 1)}\frac{\partial\chi}{\partial y}, \qquad \frac{\partial\psi}{\partial y} = \frac{1}{k_0(1 - y^2)}\frac{\partial\chi}{\partial x}.$$

In this case, we obtain the Reissner–Nordstrom solution

$$f = 1 - \frac{2m}{r} + \frac{e^2}{r^2}, \quad A_3' = A_4 = -\frac{e}{r} \tag{4.48}$$

$(x = \frac{r-m}{k_0}, \; y = \cos\theta)$, which can describe a charged non-rotating black hole.

Equations (4.44) and (4.46) take the form

$$\frac{\partial\theta}{\partial r} = \frac{\partial\theta}{\partial\psi} = 0, \tag{4.49}$$

which gives us the solution $\theta = c_0$.

3.

$$c_0 = 0, \quad b_0 = -d_0, \quad a_0 = -2md_0,$$

$$k = \frac{m}{p}, \quad \sigma = \frac{qm}{p}, \quad p^2 + q^2 = 1,$$

$$\chi = k(x + p + qy), \quad \psi = \frac{1}{2}\ln\left[\frac{1+y}{1-y}\left(\frac{x+1}{x-1}\right)^q\right], \tag{4.50}$$

and we obtain the Gutsunaev–Elsgolts solution [25]:

$$f = \frac{4m^2(r^2 - 2mr - \sigma^2)M^2}{[(r + \sigma\cos\theta)^2 - (r^2 - 2mr - \sigma^2)M^2]}, \tag{4.51}$$

$$mA_3 = (\cos\theta - \tanh\psi)(r + \sigma\cos\theta)^2 + \sigma(r + \sigma\cos\theta + m)\sin^2\theta, \tag{4.52}$$

$$A_3' = A_4 = 1 - \frac{2m(r + \sigma\cos\theta)}{(r + \sigma\cos\theta)^2 - (r^2 - 2mr - \sigma^2)M^2}, \tag{4.53}$$

$$M = \sin^2\frac{\theta}{2}\left[\frac{\sqrt{1-q^2}\,(r-m) - m}{\sqrt{1-q^2}\,(r-m) + m}\right]^{q/2}$$

$$+ \cos^2\frac{\theta}{2}\left[\frac{\sqrt{1-q^2}\,(r-m) + m}{\sqrt{1-q^2}\,(r-m) - m}\right]^{q/2}, \tag{4.54}$$

$$\tanh\psi = 2\left[1 + \frac{1-\cos\theta}{1+\cos\theta}\left(\frac{r-m-k}{r-m+k}\right)^q\right]^{-1}. \tag{4.55}$$

Equations (4.44) and (4.45) in this case are

$$\frac{dr}{d\theta} = \frac{A}{B}, \quad \frac{d\theta}{d\psi} = 0, \tag{4.56}$$

where

$$A = \frac{\partial}{\partial\theta}\left[(r + q\cos\theta)^2 \tanh\psi\right]$$

$$- \left[r^2 - 2mr - q^2 m^2\right]\sin^2\theta + 2(r - m)(r + q\cos\theta)\sin\theta,$$

$$B = \frac{\partial}{\partial r}\left[(r + q\cos\theta)^2 \tanh\psi\right] - \left[2r + q\cos\theta\right]\cos\theta - q.$$

The solution of equations (4.56) has the form

$$(r^2 + \sigma^2)\cos\theta + \sigma[r + m + (r - m)\cos\theta]$$

$$- (r + \sigma\cos\theta)^2 \tanh\psi = C_0. \tag{4.57}$$

The asymptotic behavior of the potential A_3 (or A_3') as $r \to \infty$ is

$$A_3 \sim -\frac{2\sigma m}{3}\frac{\sin^2\theta}{r}, \quad A_3' \sim -\frac{2\sigma m}{3}\frac{\cos\theta}{r^2} \tag{4.58}$$

shows that the solution describes the gravitational field of a star endowed with a magnetic dipole moment. For the asymptotic behavior of (4.51)–(4.55), when $r \to \infty$, we have the following equations of the lines of force:

$$\frac{dr}{d\theta} = 2r\cot\theta, \quad \frac{d\theta}{d\varphi} = 0, \tag{4.59}$$

where $\frac{\sin^2\theta}{r} = C_0$ represents their solution describing the family of lines of force for magnetic dipole in classical electrodynamics.

4.6 One-Stationary Euclidon Solution in a Static Electrovacuum Gravitational Field

For equations (4.4), (4.7), it is straightforward to obtain a class of solutions analogous to the Lewis class:

$$u = \rho\frac{\cosh(\psi + U_0)}{C_1 \sinh U_0},$$

$$A_3' = \frac{1}{C_1 \sinh U_0}\frac{\partial\chi}{\partial z} + C_2, \quad A_3 = C_1\frac{\cosh\psi}{\cosh(\psi + U_0)} + C_3, \tag{4.60}$$

where C_1, C_2, C_3 and U_0 are arbitrary constants. The functions $\psi(\rho, z) = \frac{1}{\rho}\frac{\partial \chi}{\partial \rho}$ and $\chi(\rho, z)$ satisfy the following linear equations:

$$\frac{\partial^2 \psi}{\partial \rho^2} + \frac{1}{\rho}\frac{\partial \psi}{\partial \rho} + \frac{\partial^2 \psi}{\partial z^2} = 0,$$

$$\frac{\partial^2 \chi}{\partial \rho^2} - \frac{1}{\rho}\frac{\partial \chi}{\partial \rho} + \frac{\partial^2 \chi}{\partial z^2} = 0.$$

If we set

$$\chi = \chi_1 = \int \sqrt{\rho^2 + (z - z_1)^2}\, dz, \quad z_1 = \text{const},$$

then

$$\frac{\partial \chi_1}{\partial z} = \sqrt{\rho^2 + (z - z_1)^2}, \quad \psi = \psi_1 = \frac{1}{\rho}\frac{\partial \chi_1}{\partial \rho}$$

$$- \ln \rho + \ln \left[(z - z_1) + \sqrt{\rho^2 + (z - z_1)^2}\right].$$

The expressions (4.60) in this particular case take the form

$$u = \frac{1}{C_1}\left[(z - z_1) + \sqrt{\rho^2 + (z - z_1)^2}\,\coth U_0\right],$$

$$A_3' = \frac{1}{C_1}\frac{\sqrt{\rho^2 + (z - z_1)^2}}{\sinh U_0} + C_2, \tag{4.61}$$

$$A_3 = C_1 \frac{\sqrt{\rho^2 + (z - z_1)^2}}{z - z_1 + \sqrt{\rho^2 + (z - z_1)^2}\,\coth U_0 \sinh U_0} + C_3.$$

Let us find the superposition of solution (4.61), which is analogous to the stationary Euclidean solution, with an arbitrary axially symmetric Einstein–Maxwell field $(u^0, A_3'^0, A_3^0)$.

Using the method of variation of constants, let us perform the substitutions

$$C_1 \to u_0(\rho, z), \quad C_2 \to A_3^0(\rho, z), \quad C_3 \to A_3'^0(\rho, z), \quad U_0 \to U(\rho, z).$$

The resulting field in this case takes the form

$$u = \frac{1}{u_0}\left[(z - z_1) + \sqrt{\rho^2 + (z - z_1)^2}\,\coth U(\rho, z)\right],$$

$$A_3' = \frac{1}{u_0}\frac{\sqrt{\rho^2 + (z - z_1)^2}}{\sinh U(\rho, z)} + A_3^0(\rho, z),$$

$$A_3 = u_0\frac{\sqrt{\rho^2 + (z - z_1)^2}}{z - z_1 + \sqrt{\rho^2 + (z - z_1)^2}\,\coth U(\rho, z)\sinh U(\rho, z)}$$

$$+ A_3'^0(\rho, z). \tag{4.62}$$

Substituting expressions (4.62) into equations (4.4), (4.7) gives a system of two nonlinear equations for the function $U(\rho, z)$ [33]:

$$\sqrt{\rho^2 + (z - z_1)^2}\,\frac{\partial U}{\partial \rho}$$

$$= (z - z_1)\frac{1}{u_0}\frac{\partial u_0}{\partial \rho} + \rho\frac{1}{u_0}\frac{\partial u_0}{\partial z} + \rho\frac{\cosh U}{u_0}\frac{\partial A_3'^0}{\partial z}$$

$$+ \left[(z - z_1)\cosh U + \sqrt{\rho^2 + (z - z_1)^2}\,\sinh U\right]\frac{1}{u_0}\frac{\partial A_3'^0}{\partial \rho},$$

$$\sqrt{\rho^2 + (z - z_1)^2}\,\frac{\partial U}{\partial z}$$

$$= -\rho\frac{1}{u_0}\frac{\partial u_0}{\partial \rho} + (z - z_1)\frac{1}{u_0}\frac{\partial u_0}{\partial z} - \rho\cosh U\frac{1}{u_0}\frac{\partial A_3'^0}{\partial \rho}$$

$$+ [(z - z_1)\cosh U + \sqrt{\rho^2 + (z - z_1)^2}\,\sinh U]\frac{1}{u_0}\frac{\partial A_3'^0}{\partial z}. \tag{4.63}$$

The integrability condition for (4.63) is satisfied. The substitution $U = \ln\frac{a}{b}$ leads to a system of linear equations for the functions $a(\rho, z)$ and $b(\rho, z)$:

$$2u_0\frac{\partial a}{\partial \rho} = a\left(\hat{L}_1 - \frac{\partial}{\partial \rho}\right)u_0 + b\left(\hat{L}_1 - \frac{\partial}{\partial \rho}\right)A_3'^0,$$

$$2u_0\frac{\partial a}{\partial z} = -a\left(\hat{L}_2 + \frac{\partial}{\partial z}\right)u_0 - b\left(\hat{L}_2 + \frac{\partial}{\partial z}\right)A_3'^0,$$

$$2u_0\frac{\partial b}{\partial \rho} = -b\left(\hat{L}_1 + \frac{\partial}{\partial \rho}\right)u_0 - a\left(\hat{L}_1 + \frac{\partial}{\partial \rho}\right)A_3'^0,$$

$$2u_0\frac{\partial b}{\partial z} = b\left(\hat{L}_2 - \frac{\partial}{\partial z}\right)u_0 + a\left(\hat{L}_2 - \frac{\partial}{\partial z}\right)A_3'^0,$$

where

$$\sqrt{\rho^2 + (z - z_1)^2}\,\hat{L}_1 = (z - z_1)\frac{\partial}{\partial \rho} + \rho\frac{\partial}{\partial z},$$

$$\sqrt{\rho^2 + (z - z_1)^2}\,\hat{L}_2 = \rho\frac{\partial}{\partial \rho} - (z - z_1)\frac{\partial}{\partial z}.$$

Let us rewrite the basic relationships in the prolate ellipsoidal coordinates (x, y):

$$\rho = k_0\sqrt{(x^2 - 1)(1 - y^2)}, \quad z = k_0 xy,$$

$$\frac{\partial A_3^0}{\partial x} = \frac{k_0(1 - y^2)}{u_0^2}\frac{\partial A_3'^0}{\partial y}, \quad \frac{\partial A_3^0}{\partial y} = \frac{-k_0(x^2 - 1)}{u_0^2}\frac{\partial A_3'^0}{\partial x}.$$

Thus,

$$u = k_0\frac{(xy - 1) + (x - y)\coth U}{u_0}, \quad A_3' = \frac{k_0(x - y)}{u_0\cosh U} + A_3^0,$$

$$A_3 = \frac{u_0(x - y)}{(xy - 1) + (x - y)\coth U} \cdot \frac{1}{\sinh U} + A_3'^0 \tag{4.64}$$

and

$$\frac{\partial U}{\partial x} = \frac{(xy - 1)}{(x - y)}\frac{1}{u_0}\frac{\partial u_0}{\partial x} + \frac{(1 - y^2)}{(x - y)}\frac{1}{u_0}\frac{\partial u_0}{\partial y} + \left[\sinh U + \frac{(xy - 1)}{(x - y)}\cosh U\right]$$

$$\times \frac{1}{u_0}\frac{\partial A_3'^0}{\partial x} + \left(\frac{1 - y^2}{x - y}\right)\frac{\cosh U}{u_0}\frac{\partial A_3'^0}{\partial y},$$

$$\frac{\partial U}{\partial y} = -\frac{(x^2 - 1)}{(x - y)}\frac{1}{u_0}\frac{\partial u_0}{\partial x} + \frac{(xy - 1)}{(x - y)}\frac{1}{u_0}\frac{\partial u_0}{\partial y} + \left[\sinh U + \frac{(xy - 1)}{(x - y)}\cosh U\right]$$

$$\times \frac{1}{u_0}\frac{\partial A_3'^0}{\partial y} - \left(\frac{x^2 - 1}{x - y}\right)\frac{\cosh U}{u_0}\frac{\partial A_3'^0}{\partial x}. \tag{4.65}$$

Let us consider an application of this method. As the external field, let us choose

$$u_0 = (x + 1)^\delta(x - 1)^{1-\delta}(1 + y)^{1-\gamma}(1 - y)^\gamma, \quad A_3^0 = A_3'^0 = 0, \tag{4.66}$$

where δ and γ are arbitrary constants.

In this case, we have

$$\lim_{U\to\infty} f = \lim_{U\to\infty} u^2 = \left(\frac{x-1}{x+1}\right)^{2\delta}\left(\frac{1+y}{1-y}\right)^{2\gamma}. \tag{4.67}$$

It is easy to see that expression (4.67) for f satisfies the static Einstein field equations:

$$\frac{\partial}{\partial x}\left((x^2-1)\frac{\partial \ln f}{\partial x}\right) + \frac{\partial}{\partial y}\left((1-y^2)\frac{\partial \ln f}{\partial y}\right) = 0. \tag{4.68}$$

In the case when $\gamma = 0$, expression (4.67) reduces to the Zipoy solution.

After integrating (4.65), we find the unknown function $U(x,y)$:

$$U(x,y) = \ln\left[\frac{(x-y)^{2(1-\gamma-\delta)}}{\beta}\frac{(x^2-1)^\delta(1-y^2)^\gamma}{(x-1)(1+y)}\right], \tag{4.69}$$

where β is an integration constant. In the limiting case $\beta \to 0$, the expression

$$f = u^2 = \left[\frac{xy-1+(x-y)\coth U}{u_0}\right]^2$$

reduces to (4.67).

Substituting (4.66) and (4.69) into (4.64) leads to the solution

$$\tilde{u} = \left(\frac{x-1}{x+1}\right)^\delta\left(\frac{1+y}{1-y}\right)^\gamma\frac{\tilde{A}}{\tilde{B}}, \quad \tilde{A}'_3 = \frac{2\beta\tilde{C}}{\tilde{B}}, \quad \tilde{A}_3 = \frac{2k_0\beta\tilde{D}}{\tilde{A}}, \tag{4.70}$$

where

$$\tilde{A} = (x-y)^{4(1-\gamma-\delta)} + \beta^2(x^2-1)^{-2\delta+1}(1-y^2)^{-2\gamma+1},$$

$$\tilde{B} = (x-y)^{4(1-\gamma-\delta)} - \beta^2(x+1)^{-2\delta}(x-1)^{-2\delta+2}(1+y)^{-2\gamma+2}$$
$$\times (1-y)^{-2\gamma},$$

$$\tilde{C} = (x-y)^{1+2(1-\gamma-\delta)}(x+1)^{-2\delta}(1-y)^{-2\gamma},$$

$$\tilde{D} = (x-y)^{1+2(1-\gamma-\delta)}(x-1)^{-2\delta+1}(1+y)^{-2\gamma+1}.$$

Apply the symmetry theorem to the obtained solution (4.70) to derive a more general solution

$$u = \frac{(-ad+bc)\tilde{u}}{(c-d\tilde{A}'_3)^2 - d^2\tilde{u}^2}, \qquad A'_3 = \frac{-ac+(ad+bc)\tilde{A}'_3 + bd(\tilde{u}^2 - \tilde{A}'^2_3)}{(c-d\tilde{A}'_3)^2 - d^2\tilde{u}^2}.$$

$$(4.71)$$

Introduce the following notations:

$$P^+ = \frac{c^2 - d^2}{c^2 + d^2}, \quad Q^+ = \frac{2cd}{c^2 + d^2}, \quad (P^+)^2 + (Q^+)^2 = 1,$$

$$P^- = \frac{ac - bd}{ac + bd}, \quad Q^- = \frac{ad + bc}{ac + bd},$$

$$R^- = \frac{-ad + bc}{ac + bd}, \quad (R^-)^2 + 1 = (P^-)^2 + (Q^-)^2.$$

$$(4.72)$$

Thus, the final solution (4.71) takes the form

$$u = AB_+^{-1}, \qquad A'_3 = B_- B_+^{-1},$$

$$R^- A_3 = 2k_0(K_+ - K_- + L)M^{-1} + 2k_0 Q^+(\delta y - \gamma x),$$

$$A = R^- u_0 \left[xy - 1 + (x-y)\coth U \right],$$

$$U = \ln\left[\frac{1}{\beta}(x+1)^\delta(x-1)^{\delta-1} \cdot (1+y)^{\gamma-1}(1-y)^\gamma(x-y)^{2(1-\gamma-\delta)} \right],$$

$$2B_\pm = (1+P^\pm)u_0^2 - 2Q^\pm u_0(x-y)\sinh^{-1} U - (1-P^\pm)W,$$

$$u_0 = (x+1)^\delta(x-1)^{1-\delta}(1+y)^{1-\gamma}(1-y)^\gamma,$$

$$2K_\pm = \beta(x-y)^{1+2(1-\gamma-\delta)} \cdot (x\pm 1)^{-2\delta+1}(1\mp y)^{-2\gamma+1}(1\mp P^+),$$

$$L = \beta^2 Q^+(x-y)(x^2-1)^{-2\delta+1} \cdot (1-y^2)^{-2\gamma+1},$$

$$M = (x-y)^{4(1-\gamma-\delta)} + \beta^2(x^2-1)^{-2\delta+1} \cdot (1-y^2)^{-2\gamma+1},$$

$$W = (x-y)^2 + (xy-1)^2 + 2(x-y)(xy-1)\coth U. \quad (4.73)$$

Now, let us consider the most physically interesting solutions from this series:

1. $\delta = +1/2$, $\beta(c^2 + d^2) = cd$ $(Q^+ = 2\beta)$, $\gamma = 0$.

In this case, we obtain

$$u = \frac{(c^2 - d^2)\sqrt{x^2 - 1}\left[(x - y)^2 + \beta^2(1 - y^2)\right]}{M},$$

$$A_3' = \frac{c^2 - d^2}{bc - ad} \cdot \frac{L}{M}, \tag{4.74}$$

$$A_3 = \frac{2cd(x - y) + \beta\left[c^2(1 + y) - d^2(1 - y)\right]}{(x - y)^2 + \beta^2(1 - y^2)} \cdot \frac{2k_0\beta^2(1 - y^2)}{c^2 - d^2},$$

where

$$L = (x^2 - y^2)[ac(x + 1) - bd(x - 1)] - (bc + ad)2\beta(x - y)^2$$
$$- \beta^2[ac(x - 1)(1 + y)^2 - bd(x + 1)(1 - y)^2],$$
$$M = (x - y)^2[c^2(x + 1) - d^2(x - 1)] - 2cd \cdot 2\beta(x - y)^2$$
$$- \beta^2[c(x - 1)(1 + y)^2 - d^2(x + 1)(1 - y)^2].$$

Under the coordinate transformation

$$x = \frac{r - m}{k_0}, \quad y = \cos\vartheta, \quad k_0 = m\frac{c^2 + d^2}{c^2 - d^2} = \frac{m}{P^+}, \tag{4.75}$$

the asymptotic behavior of the functions f, A_3, A_3' as $r \to \infty$ is given by

$$f = u^2 \to 1 - \frac{2m}{r}, \quad A_3 \to \frac{\mu\sin^2\vartheta}{r}, \quad A_3' \to -\frac{\mu\cos\vartheta}{r^2}. \tag{4.76}$$

Here, the parameter m is the total mass and the magnetic dipole moment μ is

$$\mu = \frac{4\beta^3 k_0^3}{m}. \tag{4.77}$$

If magnetic parameter equals zero ($\mu = 0$), solution (4.74) transitions to the Schwarzschild solution.

2. $\delta = -1/2$, $\beta(c^2 + d^2) = -cd$ $(Q^+ = -2\beta)$, $\gamma = 0$.

In this case, the solution is obtained as follows:

$$u = \frac{(c^2 - d^2)\sqrt{x^2 - 1}[(x - y)^6 + \beta^2(x^2 - 1)^2(1 - y^2)]}{D},$$

$$A_3' = \frac{c^2 - d^2}{bc - ad}\frac{E}{D}, \tag{4.78}$$

$$A_3 = \frac{k_0}{c^2 - d^2}\frac{H}{(x - y)^6 + \beta^2(x^2 - 1)^2(1 - y^2)} - \frac{2cdk_0y}{c^2 - d^2},$$

where

$$E = (x - y)^6[ac(x - 1) - bd(x + 1)] - (bc + ad)2\beta(x - y)^4(x^2 - 1)$$

$$- \beta^2(x^2 - 1)[ac(x - 1)^3(1 + y)^2 - bd(x + 1)^3(1 - y)^2],$$

$$D = (x - y)^6[c^2(x - 1) - d^2(x + 1)] - 2cd2\beta(x - y)^4(x^2 - 1)$$

$$- \beta^2(x^2 - 1)[c^2(x - 1)^3(1 + y)^2 - d^2(x + 1)^3(1 - y)^2],$$

$$H = -2\beta(x - y)^4[c^2(x - 1)^2(1 + y) - d^2(x + 1)^2(1 - y)]$$

$$+ 4cd\beta^2(x - y)(x^2 - 1)^2(1 - y^2).$$

Under the coordinate transformation

$$x = \frac{r - m}{k_0}, \quad y = \cos\vartheta, \quad k_0 = -m\frac{c^2 + d^2}{c^2 - d^2} = -\frac{m}{P^+}, \tag{4.79}$$

the asymptotic behavior of the solution is

$$f = u^2 \to 1 - \frac{2m}{r}, \quad A_3 \to -\frac{4k_0^3\beta}{m}(1 - \beta^2)\frac{\sin^2\vartheta}{r}, \quad A_3' \to \frac{\mu\cos\vartheta}{r^2}. \tag{4.80}$$

Here, the total mass m and the magnetic dipole moment μ are

$$\mu = -\frac{4k_0^2\beta}{m}(1 - \beta^2). \tag{4.81}$$

In the absence of a magnetic field, solution (4.78) reduces to the Schwarzschild solution.

4.7 One-Stationary Soliton Solution in a Static Electrovacuum Gravitational Field

One of the simplest solutions of equations (4.4), (4.7) is the soliton solution:

$$u = C_1 \frac{\rho}{\epsilon_1(z - z_1) + \sqrt{\rho^2 + (z - z_1)^2}\,\tanh U_0},$$

$$A_3' = \frac{\sqrt{\rho^2 + (z - z_1)^2}}{\epsilon_1(z - z_1) + \sqrt{\rho^2 + (z - z_1)^2}\,\tanh U_0} \frac{1}{\cosh U_0} + C_2, \quad (4.82)$$

$$A_3 = \epsilon_1 \frac{\sqrt{\rho^2 + (z - z_1)^2}}{C_1 \cosh U_0} + C_3,$$

where $\epsilon_1 = \pm 1$, C_1, C_2, C_3 and U_0 are arbitrary constants.

Then, using (4.82), one can construct the solution

$$u = \frac{1}{u^0} \cdot \frac{\rho^2}{\epsilon_1(z - z_1) + \sqrt{\rho^2 + (z - z_1)^2}\,\tanh U},$$

$$A_3' = \frac{\rho}{u^0} \cdot \frac{\sqrt{\rho^2 + (z - z_1)^2}}{\epsilon_1(z - z_1) + \sqrt{\rho^2 + (z - z_1)^2}\,\tanh U} \frac{1}{\cosh U} + A_3^0,$$

$$A_3 = \frac{u^0}{\rho} \cdot \frac{\epsilon_1 \sqrt{\rho^2 + (z - z_1)^2}}{\cosh U} + A_3'^0, \quad (4.83)$$

where $u^0(\rho, z)$, $A_3'^0(\rho, z)$ and $A_3^0(\rho, z)$ are arbitrary solutions also satisfying equations (4.4) and (4.7), i.e.,

$$u^0 \Delta u^0 = (\vec{\nabla} u^0)^2 + \frac{u^{0^4}}{\rho^2}(\vec{\nabla} A_3^0)^2,$$

$$\vec{\nabla}\left(\frac{u^{0^2}}{\rho^2} \vec{\nabla} A_3^0\right) = 0, \quad (4.84)$$

$$\frac{\partial A_3^{0'}}{\partial \rho} = \frac{u^{0^2}}{\rho} \cdot \frac{\partial A_3^0}{\partial z}, \quad \frac{\partial A_3'^0}{\partial z} = -\frac{u^{0^2}}{\rho} \cdot \frac{\partial A_3'^0}{\partial \rho}, \quad (4.85)$$

which we call the seed solution.

The unknown function $U(\rho, z)$ can be calculated from the equations

$$\sqrt{\rho^2 + (z - z_1)^2}\,\frac{\partial U}{\partial \rho}$$

$$= \epsilon_1(z - z_1)\left(\frac{1}{\rho} - \frac{1}{u^0}\frac{\partial u^0}{\partial \rho}\right) - \frac{\epsilon_1 \rho}{u^0}\frac{\partial u^0}{\partial z} + \epsilon_1 \sinh U \cdot u^0 \frac{\partial A_3^0}{\partial z}$$

$$+ \left[\sqrt{\rho^2 + (z - z_1)^2}\cosh U + \epsilon_1(z - z_1)\sinh U\right]\frac{u^0}{\rho}\frac{\partial A_3^0}{\partial \rho},$$

$$\sqrt{\rho^2 + (z - z_1)^2}\,\frac{\partial U}{\partial z}$$

$$= -\epsilon_1 \rho\left(\frac{1}{\rho} - \frac{1}{u^0}\frac{\partial u^0}{\partial \rho}\right) - \frac{\epsilon_1(z - z_1)}{u^0}\frac{\partial u^0}{\partial z} - \epsilon_1 \sinh U \cdot u^0 \frac{\partial A_3^0}{\partial \rho}$$

$$+ \left[\sqrt{\rho^2 + (z - z_1)^2}\cosh U + \epsilon_1(z - z_1)\sinh U\right]\frac{u^0}{\rho}\frac{\partial A_3^0}{\partial z}.$$

$$(4.86)$$

Equations (4.86) are nonlinear. Yet, if one takes

$$U(\rho, z) = \ln \frac{a(\rho, z)}{b(\rho, z)},$$

then the set of equations for the functions $a(\rho, z)$ and $b(\rho, z)$ becomes linear:

$$2\frac{\partial a}{\partial \rho} = [\tilde{M}]a - [Q_-]b, \quad 2\frac{\partial b}{\partial \rho} = -[\tilde{M}]b - [Q_+]a,$$

$$2\frac{\partial a}{\partial z} = [\tilde{N}]a + [P_-]b, \quad 2\frac{\partial b}{\partial z} = -[\tilde{N}]b + [P_+]a, \qquad (4.87)$$

where

$$\sqrt{\rho^2 + (z - z_1)^2}[\tilde{M}] = \frac{\epsilon_1(z - z_1)}{\rho}u^0 \cdot \frac{\partial}{\partial \rho}\left(\frac{\rho}{u^0}\right) + \epsilon_1 u^0 \cdot \frac{\partial}{\partial z}\left(\frac{\rho}{u^0}\right),$$

$$\sqrt{\rho^2 + (z - z_1)^2}[\tilde{N}] = -\epsilon_1 u^0 \cdot \frac{\partial}{\partial \rho}\left(\frac{\rho}{u^0}\right) + \frac{\epsilon_1(z - z_1)u^0}{\rho} \cdot \frac{\partial}{\partial z}\left(\frac{\rho}{u^0}\right),$$

$$\sqrt{\rho^2 + (z - z_1)^2}\,[P_\pm]$$

$$= -\epsilon_1 u^0 \cdot \frac{\partial A_3^0}{\partial \rho} + \left[\epsilon_1(z - z_1) \pm \sqrt{\rho^2 + (z - z_1)^2}\,\right] \frac{u^0}{\rho} \cdot \frac{\partial A_3^0}{\partial z},$$

$$\sqrt{\rho^2 + (z - z_1)^2}\,[Q_\pm]$$

$$= -\epsilon_1 u^0 \cdot \frac{\partial A_3^0}{\partial z} - \left[\epsilon_1(z - z_1) \pm \sqrt{\rho^2 + (z - z_1)^2}\,\right] \frac{u^0}{\rho} \cdot \frac{\partial A_3^0}{\partial \rho}.$$

It is sometimes convenient to pass to the prolate ellipsoidal coordinates (x, y), connected with the Weyl canonical coordinates (ρ, z) by the relations

$$\rho = k_0 \sqrt{(x^2 - 1)(1 - y^2)}, \quad z = k_0 x y \tag{4.88}$$

(k_0 is a real constant). In this case, (4.85), (4.83) and (4.86) become for $k_0 = z_1$

$$\frac{\partial A_3'^0}{\partial x} = \frac{u^{0^2}}{k_0(x^2 - 1)} \cdot \frac{\partial A_3^0}{\partial y},$$

$$\frac{\partial A_3'^0}{\partial y} = -\frac{u^{0^2}}{k_0(1 - y^2)} \cdot \frac{\partial A_3^0}{\partial x}, \tag{4.89}$$

$$u = \frac{k_0(x^2 - 1)(1 - y^2)}{\epsilon_1(xy - 1) + (x - y)\tanh U} \cdot \frac{1}{u^0},$$

$$A_3' = \frac{k_0 \sqrt{(x^2 - 1)(1 - y^2)}(x - y)}{\epsilon_1(xy - 1) + (x - y)\tanh U} \cdot \frac{1}{\cosh U}\frac{1}{u^0} + A_3^0,$$

$$A_3 = \frac{\epsilon_1(x - y)}{\sqrt{(x^2 - 1)(1 - y^2)}} \cdot \frac{u^0}{\cosh U} + A_3'^0 \tag{4.90}$$

and

$$\frac{\partial U}{\partial x} = -\frac{\epsilon_1}{x^2 - 1} - \frac{\epsilon_1(xy - 1)}{x - y} \cdot \frac{1}{u^0} \cdot \frac{\partial u^0}{\partial x} - \frac{\epsilon_1(1 - y^2)}{x - y} \cdot \frac{1}{u^0} \cdot \frac{\partial u^0}{\partial y}$$

$$+ \left[\cosh U + \frac{\epsilon_1(xy - 1)}{x - y}\sinh U\right] \frac{u^0}{k_0 \sqrt{(x^2 - 1)(1 - y^2)}}$$

$$\cdot \frac{\partial A_3^0}{\partial x} + \frac{\epsilon_1(1-y^2)}{x-y}\sinh U \frac{u^0}{k_0\sqrt{(x^2-1)(1-y^2)}}\frac{\partial A_3^0}{\partial y},$$

$$\frac{\partial U}{\partial y} = -\frac{\epsilon_1}{1-y^2} + \frac{\epsilon_1(x^2-1)}{x-y}\cdot\frac{1}{u^0}\cdot\frac{\partial u^0}{\partial x}$$

$$-\frac{\epsilon_1(xy-1)}{x-y}\cdot\frac{1}{u^0}\cdot\frac{\partial u^0}{\partial y} + \left[\cosh U + \frac{\epsilon_1(xy-1)}{x-y}\sinh U\right]$$

$$\times \frac{u^0}{k_0\sqrt{(x^2-1)(1-y^2)}}\cdot\frac{\partial A_3^0}{\partial y} - \frac{\epsilon_1(x^2-1)}{x-y}$$

$$\cdot\sinh U \frac{u^0}{k_0\sqrt{(x^2-1)(1-y^2)}}\frac{\partial A_3^0}{\partial x}. \tag{4.91}$$

Let us consider the seed solution $(u^0, A_3^{\prime 0}, A_3^0)$ of the form

$$u^0 = (x+1)^{1+\delta}(x-1)^{-\delta}(1-y), \quad A_3^{\prime 0} = A_3^0 = 0. \tag{4.92}$$

In this case, from (4.90) with (4.91), it follows ($\epsilon_1 = 1, k_0 = 1$)

$$\tilde{u} = \frac{1}{(x+1)^{1+\delta}(x-1)^{-\delta}(1-y)}\frac{(x^2-1)(1-y^2)}{xy-1+(x-y)\tanh U},$$

$$\tilde{A}_3 = \frac{(x+1)^{1+\delta}(x-1)^{-\delta}(1-y)}{\sqrt{(x^2-1)(1-y^2)}}\frac{x-y}{\cosh U}, \tag{4.93}$$

$$\tilde{A}_3' = \frac{\sqrt{(x^2-1)(1-y^2)}}{(x+1)^{1+\delta}(x-1)^{-\delta}(1-y)}\frac{x-y}{xy-1+(x-y)\tanh U}\frac{1}{\cosh U}$$

and

$$\frac{\partial U}{\partial x} = -\frac{\delta+\frac{1}{2}}{x-1} - \frac{\delta+\frac{1}{2}}{x+1} + \frac{2(1+\delta)}{x-y},$$

$$\frac{\partial U}{\partial y} = -\frac{\frac{1}{2}}{1+y} + \frac{\frac{1}{2}}{1-y} - \frac{2(1+\delta)}{x-y}. \tag{4.94}$$

Integration of (4.94) gives

$$U = \ln\left[(x-y)^{2(1+\delta)}(x^2-1)^{-(\delta+\frac{1}{2})}(1-y^2)^{-\frac{1}{2}}\right] - \ln\beta, \tag{4.95}$$

where β is an integration constant,

$$\sinh U = \frac{(x-y)^{4(1+\delta)}(x^2-1)^{-2(\delta+\frac{1}{2})} - \beta^2(1-y^2)}{2\beta(x-y)^{2(1+\delta)}(x^2-1)^{-(\delta+\frac{1}{2})}(1-y^2)^{1/2}},$$

$$\cosh U = \frac{(x-y)^{4(1+\delta)}(x^2-1)^{-2(\delta+\frac{1}{2})} + \beta^2(1-y^2)}{2\beta(x-y)^{2(1+\delta)}(x^2-1)^{-(\delta+\frac{1}{2})}(1-y^2)^{1/2}}.$$

Note that if $\tilde{\varepsilon}_1 = \tilde{u} + \tilde{A}'_3$ and $\tilde{\varepsilon}_2 = \tilde{u} - \tilde{A}'_3$ form a solution of equations (4.11), one can construct another solution [32]

$$\varepsilon_1 = u + A'_3 = \frac{a + b\tilde{\varepsilon}_1}{c + d\tilde{\varepsilon}_1},$$

$$\varepsilon_2 = u - A'_3 = \frac{-a + b\tilde{\varepsilon}_2}{c - d\tilde{\varepsilon}_2},$$

where a, b, c, d are arbitrary real constants. Let us consider the physically most interesting solutions arising from this series.

1. $\delta = -\frac{1}{2}$, $\beta(c^2 + d^2) = cd$, $x \to -x$, $y \to -y$.

In this case, one can construct a solution of the form

$$u = (c^2 - d^2)\sqrt{x^2 - 1} \cdot \left[(x-y)^2 + \beta^2(1-y^2)\right] \cdot M^{-1},$$

$$A'_3 = \frac{c^2 - d^2}{bc - ad} \cdot L \cdot M^{-1},$$

$$A_3 = \frac{2cd(x-y) + \beta\left[c^2(1+y) - d^2(1-y)\right]}{(x-y)^2 + \beta^2(1-y^2)} \times \frac{2k\beta^2(1-y^2)}{c^2 - d^2},$$

$$\tag{4.96}$$

where

$$L = (x-y)^2\left[ac(x+1) - bd(x-1)\right] - (bc + ad)2\beta(x-y)^2$$
$$- \beta^2\left[ac(x-1)(1+y)^2 - bd(x+1)(1-y)^2\right],$$

$$M = (x-y)^2\left[c^2(x+1)^2 - d^2(x-1)^2\right] - 2cd \cdot 2\beta(x-y)^2$$
$$- \beta^2[c^2(x-1)(1+y)^2 - d^2(x+1)(1-y)^2].$$

With the coordinate transformation

$$x = \frac{r - m}{k}, \quad y = \cos\theta, \quad k = \frac{m(c^2 + d^2)}{c^2 - d^2},$$

the asymptotic behavior of the functions f, A_3 and A_3' as $r \to \infty$ is

$$f = u^2 \to 1 - \frac{2m}{r},$$

$$A_3 \to \frac{4\beta^3 k^3}{m} \cdot \frac{\sin^2 \theta}{r},$$

$$A_3' \to -\frac{4\beta^3 k^3}{m} \cdot \frac{\cos \theta}{r^2}.$$

The parameter m is the total mass and the magnetic dipole moment μ is given by

$$\mu = \frac{4\beta^3 k^3}{m}. \tag{4.97}$$

When the magnetic parameter is equal to zero ($\beta = 0$), (4.96) reduce to the Schwarzschild solution.

2. $\delta = \frac{1}{2}$, $\beta(c^2 + d^2) = -cd$, $x \to -x$, $y \to -y$:

$$u = (c^2 - d^2)\sqrt{x^2 - 1}\left[(x - y)^6 + \beta^2(x^2 - 1)^2(1 - y^2)\right] \cdot D^{-1},$$

$$A_3' = \frac{c^2 - d^2}{bc - ad} \cdot E \cdot D^{-1},$$

$$A_3 = \frac{k}{c^2 - d^2} \frac{H}{(x - y)^6 + \beta^2(x^2 - 1)^2(1 - y^2)} - \frac{2cdky}{c^2 - d^2}, \tag{4.98}$$

where

$$E = (x - y)^6 \left[ac(x - 1) - bd(x + 1)\right]$$

$$- (bc + ad)2\beta(x - y)^4(x^2 - 1)$$

$$- \beta^2(x^2 - 1)\left[ac(x - 1)^3(1 + y)^2 - bd(x + 1)^3(1 - y)^2\right],$$

$$D = (x - y)^6 \left[c^2(x - 1) - d^2(x + 1)\right]$$

$$- 2cd \cdot 2\beta(x - y)^4(x^2 - 1)$$

$$- \beta^2(x^2 - 1)\left[c^2(x - 1)^3(1 + y)^2 - d^2(x + 1)^3(1 - y)^2\right],$$

$$H = -2\beta(x - y)^4 \left[c^2(x - 1)^2(1 + y) - d^2(x + 1)^2(1 - y)\right]$$

$$+ 4cd \cdot \beta^2(x - y)(x^2 - 1)^2(1 - y^2).$$

Carrying out the coordinate transformation

$$x = \frac{r - m}{k}, \quad y = \cos\theta, \quad k = -\frac{m(c^2 + d^2)}{c^2 - d^2}$$

in the above formulae (4.98), we arrive at the following asymptotic behavior of the functions f, A_3 and A_3' as $r \to \infty$:

$$f = u^2 \to 1 - \frac{2m}{r},$$

$$A_3 \to -\frac{4\beta k^3}{m}(1 - \beta^2) \cdot \frac{\sin^2\theta}{r},$$

$$A_3' \to \frac{4\beta k^3}{m}(1 - \beta^2) \cdot \frac{\cos\theta}{r^2}.$$

It follows from these relations that m is the total mass and the magnetic dipole moment μ is connected with β as

$$\mu = -\frac{4k^3\beta}{m}(1 - \beta^2). \tag{4.99}$$

In the case of a vanishing magnetic field ($\beta = 0$), (4.98) also reduce to the Schwarzschild solution.

The solutions (4.74), (4.78) and (4.96), (4.98) turned out to be the same since in the case of a static external field, the Euclidon method and the soliton method coincide.

4.8 Superposition of the Bonnor Solution with an Arbitrary Einstein–Maxwell Fields

In section [19], we have considered the superposition of the Kerr solution with an arbitrary stationary Einstein vacuum field, and we could see that it allowed us to obtain a number of physically interesting solutions. One may suppose that a similar procedure of the nonlinear superposition carried out for the static Einstein–Maxwell fields would also lead to physically valuable metrics.

The forms of the superposition formulae (3.50) and (3.51) indicate us a possibility of rewriting them for the case of the static electrovacuum fields with the aid of the Bonnor theorem. For this purpose, it is better to use instead of (3.50) a single relation for a complex function $\varepsilon = f + i\Phi$, i.e.,

$$\varepsilon = \varepsilon^0 - \frac{(\varepsilon^0 + \varepsilon^{0*})(1 - ia)(1 - ib)}{x(1 + ab) + iy(a - b) + (1 - ia)(1 - ib)},\qquad (4.100)$$

where $\varepsilon^0 = f^0 + i\Phi^0$ defines a known solution of the Ernst equation.

Then, we can carry out in (4.100) and in the relation for ε^* the complex continuation of the functions a and b, coming to the real potentials ε_1 and ε_2 of the form

$$\varepsilon_1 = \varepsilon_1^0 - \frac{(\varepsilon_1^0 + \varepsilon_2^0)(1 - A)(1 - B)}{x(1 - AB) + y(A - B) + (1 - A)(1 - B)},$$

$$\varepsilon_2 = \varepsilon_2^0 - \frac{(\varepsilon_1^0 + \varepsilon_2^0)(1 + A)(1 + B)}{x(1 - AB) + y(B - A) + (1 + A)(1 + B)},\qquad (4.101)$$

where ε_1^0 and ε_2^0 satisfy the equations

$$(\varepsilon_1^0 + \varepsilon_2^0)\Delta\varepsilon_1^0 = 2(\vec{\nabla}\varepsilon_1^0)^2, \quad (\varepsilon_1^0 + \varepsilon_2^0)\Delta\varepsilon_2^0 = 2(\vec{\nabla}\varepsilon_2^0)^2. \qquad (4.102)$$

The first-order differential equations for the functions A and B which originate respectively from a and b are derivable from (3.51) and have the form

$$A_{,x} = -\frac{1}{2(x+y)u^0}\{[(xy+1)u_{,x}^0 + (1-y^2)u_{,y}^0]2A$$

$$+ [(xy+1)\Phi_{,x}^0 + (1-y^2)\Phi_{,y}^0][(1-A^2) + (x+y)(1+A^2)\Phi_{,x}^0]\},$$

$$A_{,y} = -\frac{1}{2(x-y)u^0}\{[-(x^2-1)u_{,x}^0 + (xy+1)u_{,y}^0]2A$$

$$+ [-(x^2-1)\Phi_{,x}^0 + (xy+1)\Phi_{,y}^0][(1-A^2) + (x+y)$$

$$\times (1+A^2)\Phi_{,y}^0]\}, \qquad (4.103)$$

$$B_{,x} = \frac{1}{2(x-y)u^0}\{[(xy-1)u_{,x}^0 + (1-y^2)u_{,y}^0]2B$$

$$+ [(xy-1)\Phi_{,x}^0 + (1-y^2)\Phi_{,y}^0][(1-B^2) - (x-y)(1+B^2)\Phi_{,x}^0]\},$$

$$B_{,y} = \frac{1}{2(x-y)u^0}\{[-(x^2-1)u_{,x}^{\;0} +(xy-1)u_{,y}^{\;0}]2B$$

$$+ [-(x^2-1)\Phi_{,x}^{\;0} +(xy-1)\Phi_{,y}^{\;0}][(1-B^2)-(x-y)$$

$$\times (1+B^2)\Phi_{,y}^{\;0}]\},$$

where

$$u^* \equiv \sqrt{f^0} = \frac{1}{2}(\varepsilon_1^0 + \varepsilon_2^0); \quad \Phi^0 = \frac{1}{2}(\varepsilon_1^0 - \varepsilon_2^0), \qquad (4.104)$$

and the function Φ^0 is either the electric potential A_4^0 (in the case of electrostatics) or the magnetic twist potential $A_3^{0'}$ (in the case of magnetostatics).

Formulae (4.101)–(4.104) describe the superposition of the Bonnor solution with any static axisymmetric Einstein–Maxwell field.

Since equations (4.103) are the Riccati equations, their integration in general case is a very difficult problem. So, we turn to examination of the superposition formulae corresponding to $\Phi^0 = 0$, i.e., we choose as "seed" metric an arbitrary vacuum Weyl solution. In this case, the relations (4.101), (4.102) take the form

$$\varepsilon_1 = e^{\psi}\left[1 - \frac{2(1-A)(1-B)}{x(1-AB)+y(A-B)+(1-A)(1-B)}\right],$$

$$\varepsilon_2 = e^{\psi}\left[1 - \frac{2(1+A)(1+B)}{x(1-AB)+y(B-A)+(1+A)(1+B)}\right]. \qquad (4.105)$$

and

$$A_{,x} = -A(x+y)^{-1}[(xy+1)\psi_{,x} +(1-y^2)\psi_{,y}],$$
$$A_{,y} = -A(x+y)^{-1}[-(x^2-1)\psi_{,x} +(xy+1)\psi_{,y}],$$
$$B_{,x} = B(x-y)^{-1}[(xy-1)\psi_{,x} +(1-y^2)\psi_{,y}],$$
$$B_{,y} = B(x-y)^{-1}[-(x^2-1)\psi_{,x} +(xy-1)\psi_{,y}],$$

$$(4.106)$$

the function ψ satisfying the equation $\Delta\psi = 0$. Now, the integration of system (4.106) for determination of the functions A and B is straightforward, and we can illustrate the use of the relations obtained by giving as an example the electrostatic solution reducing to the Schwarzschild metric in the absence of the electric field.

Choosing $B = 0$, $\psi = -\frac{1}{2}\ln[(x-1)/(x+1)]$ in the above formulae, one comes to the following electrostatic solution:

$$f = \frac{1}{4}(\varepsilon_1 + \varepsilon_2)^2 = \frac{(x^2 - 1)[(x+y)^2 + \alpha^2(1-y^2)]^2}{[(x+1)(x+y)^2 - \alpha^2(x-1)(1-y)^2]^2},$$

$$A_4 = \frac{1}{2}(\varepsilon_1 - \varepsilon_2) = -\frac{2\alpha(x+y)^2}{(x+1)(x+y)^2 - \alpha^2(x-1)(1-y)^2},$$

$$\tag{4.107}$$

$$e^{2\gamma} = \frac{x^2 - 1}{x^2 - y^2}\left[1 + \frac{\alpha^2(1-y^2)}{(x+y)^2}\right]^4,$$

where α is an integration constant.

When $\alpha = 0$ (no electric field), expressions (4.107) reduce to the Schwarzschild solution.

Solution (4.107) is the simplest example demonstrating how the superposition formulae (4.105), (4.106) can be used for obtaining an asymptotically flat electrostatic solution reducing to the Schwarzschild metric in the vacuum limit. Since solutions of this kind exhibit a definite physical interest, we dwell in the following on some other possibilities, taking into account that the equations for the potentials ε_1 and ε_2 produce either the electrostatic or magnetostatic solutions.

A possible generalization of solution (4.107) can be performed if both the functions A and B are not equal to zero. Choosing the function ψ in the form

$$\psi = -\frac{1}{2}\ln\left(\frac{x-1}{x+1}\right), \tag{4.108}$$

we find integrating (4.106) that

$$A = \frac{\alpha\sqrt{x^2 - 1}}{x+y}, \quad B = \frac{\beta\sqrt{x^2 - 1}}{x-y}, \tag{4.109}$$

where α and β are the integration constants.

The expressions for ε_1 and ε_2 are then found to be

$$\varepsilon_1 = \frac{E_1^+ + E_2^+}{E_1^- - E_2^-}, \quad \varepsilon_2 = \frac{E_1^+ - E_2^+}{E_1^- + E_2^-}, \tag{4.110}$$

where

$$E_1^{\pm} = [x^2 - y^2 - \alpha\beta(x \pm 1)^2]\sqrt{x \mp 1},$$

$$E_2^{\pm} = [\alpha(x - y)(1 \pm y) + \beta(x + y)(1 \mp y)]\sqrt{x \pm 1}.$$

The resultant expressions for f, γ and A_4 are

$$f = \frac{(x^2 - 1)C^2}{D^2}, \tag{4.111}$$

$$e^{2\gamma} = \frac{(x^2 - 1)C^4}{(1 - \alpha\beta)^8(x^2 - y^2)^9}, \quad A_4 = \frac{2E}{D},$$

where

$$
\begin{aligned}
C &\equiv [x^2 - y^2 - \alpha\beta(x^2 - 1)]^2 + (1 - y^2)[\alpha(x - y) - \beta(x + y)]^2, \\
D &\equiv (x + 1)[x^2 - y^2 - \alpha\beta(x - 1)^2]^2 \\
&\quad - (x - 1)[\alpha(x - y)(1 - y) + \beta(x + y)(1 + y)]^2, \\
E &\equiv (\alpha + \beta)(x^2 - y^2)^2 - \alpha\beta(x^2 - 1)[\alpha(x - y)^2 + \beta(x + y)^2].
\end{aligned}
\tag{4.112}
$$

It is not difficult to see that solution (4.107) follows from (4.111), (4.112) at $\alpha = 0$. With the coordinate transformation

$$kx = r - M, \quad y = \cos\theta, \quad k = M(1 - \alpha\beta)(1 + 3\alpha\beta)^{-1}, \tag{4.113}$$

where k and M are real constants, the asymptotic behavior of the functions f and A_4 is the following:

$$f = 1 - 2Mr^{-1} + O(r^{-2}),$$

$$A_4 = 2M(\alpha + \beta)(1 + 3\alpha\beta)^{-1}r^{-1} + O(r^{-2}) \tag{4.114}$$

so that M can be defined as the total mass, while the total charge Q is given by

$$Q = 2M(\alpha + \beta)(1 + 3\alpha\beta)^{-1}. \tag{4.115}$$

Therefore, solutions (4.111), (4.112) can describe the gravitational field of the Schwarzschild black hole surrounded by the axisymmetric charge distribution.

One particular case of this solution, corresponding to the choice of constants $\beta = -\alpha$, is of special interest. Indeed, under this choice of constants, the Q is equal to zero, and A_4 becomes of the order r^{-2} so that this case represents the gravitational field of a massive electric dipole.

Remembering that the formulae obtained are also valid for the magnetostatics, we find from (4.111), (4.112) at $\beta = -\alpha$ a magnetostatic solution describing the gravitational field of a mass possessing a magnetic dipole moment:

$$f = \frac{x-1}{x+1} \left\{ \frac{\left[x^2 - y^2 + \alpha^2(x^2-1)\right]^2 + 4\alpha^2 x^2(1-y^2)}{\left[x^2 - y^2 + \alpha^2(x-1)^2\right]^2 - 4\alpha^2 y^2(x^2-1)} \right\}^2 ,$$

$$A_3' = \frac{8\alpha^3 xy(x-1)}{[x^2 - y^2 + \alpha^2(x-1)^2]^2 - 4\alpha^2 y^2(x^2-1)}, \tag{4.116}$$

$$e^{2\gamma} = \frac{x^2-1}{x^2-y^2} \frac{\left\{\left[x^2 - y^2 + \alpha^2(x^2-1)\right]^2 + 4\alpha^2 x^2(1-y^2)\right\}^4}{(1+\alpha^2)^8(x^2-y^2)^8}.$$

For this solution, the magnetic component A_3 of the electromagnetic potential is

$$A_3 = \frac{4k\alpha^3(1-y^2)\left[2(1+\alpha^2)x^3 + (1-3\alpha^2)x^2 + y^2 + \alpha^2\right]}{(1+\alpha^2)\{[x^2 - y^2 + \alpha^2(x^2-1)]^2 + 4\alpha^2 x^2(1-y^2)\}}. \tag{4.117}$$

With the coordinate transformation

$$x = k^{-1}(r-m), \quad y = \cos\theta, \quad k = m(1+\alpha^2)(1-3\alpha^2), \tag{4.118}$$

the asymptotic behavior of the functions f, A_3 and γ for $r \to \infty$ is given by

$$f = 1 - 2mr^{-1} + O(r^{-3}),$$
$$e^{2\gamma} = 1 + O(r^{-2}),$$
$$A_3 = 8m^2\alpha^3(1-3\alpha^2)^{-2}r^{-1}\sin^2\theta + O(r^{-2}), \tag{4.119}$$

whence it appears that m is the total mass, while the magnetic dipole moment μ is given by

$$\mu = 8m^2\alpha^3(1-3\alpha^2)^{-2}. \tag{4.120}$$

Solutions (4.116), (4.117) are not the only possibility to describe the gravitational field of a massive magnetic dipole. Now, we use formulae (4.105), (4.106) for construction of another asymptotically flat algebraic solution for a mass with a magnetic dipole moment. Choosing ψ in the form

$$\psi = \left(\frac{x-1}{x+1}\right)^{-3/2}, \tag{4.121}$$

we find integrating (4.106)

$$A = \alpha(x^2-1)^{\frac{3}{2}}(x+y)^{-3}, \quad B = \beta(x^2-1)^{\frac{3}{2}}(x-y)^{-3}. \tag{4.122}$$

Then, setting $\beta = -\alpha$ and taking into account that once ε_1 and ε_2 satisfy the equations (4.11), then the functions $\tilde{\varepsilon}_1 = 1/\varepsilon_1$, $\tilde{\varepsilon}_2 = 1/\varepsilon_2$ also satisfy them, we come to the solution

$$\tilde{\varepsilon}_1 = \varepsilon_1^{-1}\left(\frac{x-1}{x+1}\right)^{\frac{1}{2}}\frac{K^{(-)}}{L^{(-)}}, \quad \tilde{\varepsilon}_2 = \varepsilon_2^{-1}\left(\frac{x-1}{x+1}\right)^{\frac{1}{2}}\frac{K^{(+)}}{L^{(+)}}, \tag{4.123}$$

where

$$\begin{aligned}
K^{(\mp)} &\equiv (x^2-y^2)^3 + \alpha^2(x-1)^2(x^2-1)^2 \\
&\quad \mp 2\alpha y(x-1)(x^3+3x^2+3xy^2+y^2)(x^2-1)^{\frac{1}{2}}, \\
L^{(\mp)} &\equiv (x^2-y^2)^3 + \alpha^2(x+1)^2(x^2-1)^2 \\
&\quad \mp 2\alpha y(x+1)(x^3-3x^2+3xy^2-y^2)(x^2-1)^{\frac{1}{2}}.
\end{aligned} \tag{4.124}$$

Then, we can find that

$$\tilde{f} = \frac{x-1}{x+1}\frac{P^2}{Q^2}, \quad e^{2\tilde{\gamma}} = \frac{x^2-1}{x^2-y^2}\frac{P^4}{(1+\alpha^2)^8(x^2-y^2)^{24}},$$

$$\tilde{A}_3' = -8\alpha xy(x-1)\left[(x^2-y^2)^4 + 2\alpha^2(x^2+y^2)(x^2-1)^3\right]Q^{-1}, \tag{4.125}$$

$$\tilde{A}_3 = -4\alpha k(1-y^2)(1+\alpha^2)^{-1}RP^{-1},$$

where we introduced

$$P \equiv [(x^2 - y^2)^3 + \alpha^2(x^2 - 1)^3]^2 + 4\alpha^2 x^2(1 - y^2)(x^2 - 1)^2(x^2 + 3y^2)^2,$$

$$Q \equiv [(x^2 - y^2)^3 + \alpha^2(x + 1)^2(x^2 - 1)^2]^2$$
$$\qquad - 4\alpha^2 y^2(x + 1)^2(x^2 - 1)(x^3 - 3x^2 + 3xy^2 - y^2)^2,$$

$$R = (x^2 - y^2)^4(2x^3 + x^2 + y^2) - \alpha^2(x + 1)^2\big[- 6x^9 + 5x^8 + 4x^7 y^2$$
$$\qquad + x^6(13y^2 + 3) - 2x^5(6y^4 + 21y^2 + 1) + x^4(40y^4 + 43y^2 - 1)$$
$$\qquad + 4x^3 y^2(2y^4 - 17y^2 - 2) - 3x^2 y^2(y^4 - 11y^2 + 2)$$
$$\qquad + y^4(1 - 2x)(y^4 + y^2 - 1)\big]$$
$$\qquad + \alpha^4(x^2 - 1)^2(x + 1)^4(4x^3 - 6x^2 + 4x + 4xy^2 - y^2 - 1).$$
$$\tag{4.126}$$

To examine the asymptotic behavior of this solution, one should use the mapping

$$x = k^{-1}(r - \tilde{m}), \quad y = \cos\theta, \quad k = \tilde{m}(1 + \alpha^2)(1 + 5\alpha^2)^{-1}. \tag{4.127}$$

Then, at $r \to \infty$, we get

$$\tilde{f} = 1 - 2\tilde{m}r^{-1} + \mathcal{O}(r^{-3}), \quad e^{2\tilde{\gamma}} = 1 + \mathcal{O}(r^{-2}),$$

$$\tilde{A}_3 = -8\alpha\tilde{m}^2(1 + 2\alpha^2)(1 + 5\alpha^2)^{-2}r^{-1}\sin^2\theta + \mathcal{O}(r^{-2}). \tag{4.128}$$

It follows from these relations that $\tilde{m}$ is the total mass and the magnetic dipole moment $\tilde{\mu}$ is given by

$$\tilde{\mu} = -8\alpha\tilde{m}^2(1 + 2\alpha^2)(1 + 5\alpha^2)^{-2}. \tag{4.129}$$

This solution, as well as the previous one, reduces to the Schwarzschild metric when the magnetic field vanishes, i.e., when $\tilde{\mu} = 0$ (at $\alpha = 0$).

Up to now, we used the superposition formulae written in the prolate ellipsoidal coordinates (x, y). At the same time, transition to the Weyl canonical coordinates (ρ, z) allows us to obtain some other solutions for a massive magnetic dipole. Introducing ρ and z

by formulae

$$\rho = z_0(x^2 - 1)^{\frac{1}{2}}(1 - y^2)^{\frac{1}{2}}, \quad z = z_0 xy, \tag{4.130}$$

we can rewrite relations (4.105) in the form

$$\varepsilon_1 = e^{\psi}\frac{r_-V + r_+W - 2z_0}{r_-V + r_+W + 2z_0}, \quad \varepsilon_2 = e^{\psi}\frac{r_-V^{-1} + r_+W^{-1} - 2z_0}{r_-V^{-1} + r_+W^{-1} + 2z_0}, \tag{4.131}$$

where

$$r_- \equiv \left[\rho^2 + (z - z_0)^2\right]^{1/2}, \quad r_+ \equiv \left[\rho^2 + (z + z_0)^2\right]^{1/2},$$
$$V = (1 + B)(1 - B)^{-1}, \quad W = (1 + A)(1 - A)^{-1}, \tag{4.132}$$

while equations (4.106) for determination of A and B are rewritten as

$$r_+A_{,\rho} = -[(z + z_0)\psi_{,\rho} + \rho\psi_{,z}]A,$$
$$r_+A_{,z} = -[-\rho\psi_{,\rho} + (z + z_0)\psi_{,z}]A,$$
$$r_-B_{,\rho} = [(z - z_0)\psi_{,\rho} + \rho\psi_{,z}]B,$$
$$r_-B_{,z} = [-\rho\psi_{,\rho} + (z - z_0)\psi_{,z}]B. \tag{4.133}$$

In order to construct solutions possessing the asymptotics of a magnetic dipole, one should multiply the functions A and B by the constants $-\alpha$ and α, respectively; while solving the equations (4.133), it is necessary to set the integration constants equal to zero. Therefore, we obtain for f and A_3' the following expressions:

$$f = e^{2\psi}(1 - M - N), \quad A_3' = e^{\psi}(N - M), \tag{4.134}$$

where

$$M = 2z_0\left(r_-\bar{V} + r_+\bar{W} + 2z_0\right)^{-1},$$
$$N = 2z_0\left(r_-\bar{V}^{-1} + r_+\bar{W}^{-1} + 2z_0\right)^{-1}, \quad \bar{V} = (1 + \alpha B)(1 - \alpha B)^{-1},$$
$$\bar{W} = (1 - \alpha A)(1 + \alpha A)^{-1}, \quad \alpha = \text{const.} \tag{4.135}$$

Consider two particular solutions arising from these relations:

(A) Choosing the function ψ in the form

$$\psi = \frac{1}{2} \ln \frac{z - Z_0 + R_-}{z + Z_0 + R_+}, \tag{4.136}$$

where Z_0 is a real constant and

$$R_- \equiv \left[\rho^2 + (z - Z_0)^2\right]^{1/2}, \quad R_+ == \left[\rho^2 + (z + Z_0)^2\right]^{1/2}, \tag{4.137}$$

one can find integrating (4.133) that

$$A^2 = [\rho^2 + (z + z_0)(z + Z_0) + r_+ R_+][\rho^2 + (z + z_0)(z - Z_0) + r_+ R_-]^{-1},$$

$$B^2 = [\rho^2 + (z - z_0)(z - Z_0) + r_- R_-][\rho^2 + (z - z_0)(z + Z_0) + r_- R_+]^{-1}. \tag{4.138}$$

Formulae (4.134) and (4.136) together with the expressions found for A and B define an exact solution, the asymptotic analysis of which is most simple in the coordinates R, ϑ ($\rho = R \sin \vartheta, z = R \cos \vartheta$). When $R \to \infty$, we have

$$f = 1 - \frac{2}{R} \left(Z_0 + \frac{2z_0(1 - \alpha^2)}{1 + \alpha^2} \right) + \mathcal{O}\left(\frac{1}{R^2}\right),$$

$$A_3' = -\frac{4\alpha z_0 \left[Z_0(1 + \alpha^2) + z_0(1 - \alpha^2)\right] \cos \vartheta}{(1 + \alpha^2)^2} \frac{\cos \vartheta}{R^2} + \mathcal{O}\left(\frac{1}{R^3}\right). \tag{4.139}$$

It follows from (4.139) that the total mass is given by

$$m = Z_0 + 2z_0(1 - \alpha^2)(1 + \alpha^2)^{-1}. \tag{4.140}$$

On the other hand, at infinity, $A_3' \to \mu r^{-2} \cos \vartheta$ (i.e., $A_3 \to \mu r^{-1} \sin^2 \vartheta$) so that

$$\mu = -4\alpha z_0 \left[(1 + \alpha^2)m + z_0(\alpha^2 - 1)\right] (1 + \alpha^2)^{-2} \tag{4.141}$$

is a magnetic dipole moment.

The case $\mu = 0$ (at $z_0 = 0$) leads to the Schwarzschild solution, while assuming $m = 0$, we arrive at the case of a massless magnetic dipole moment $\mu_0 = 4\alpha z_0(1 - \alpha^2)(1 + \alpha^2)^{-2}$ with the asymptotics

$$f = 1 + \mathcal{O}(R^{-3}), \quad A_3' = \mu_0 R^{-2} \cos \vartheta + \mathcal{O}(R^{-3}). \tag{4.142}$$

(B) If we now choose the function ψ to have the form

$$\psi = \ln\left[\left(\frac{z - Z_0 + R_-}{z + Z_0 + R_+}\right)^{1/2} \frac{z + z_0 + r_+}{z - z_0 + r_-}\right], \tag{4.143}$$

then the integration of (4.133) yields

$$A = \left(\frac{\rho^2 + (z + z_0)(z + Z_0) + r_+ R_+}{\rho^2 + (z + z_0)(z - Z_0) + r_+ R_-}\right)^{1/2} \frac{\rho^2 + z^2 - z_0^2 + r_- r_+}{2r_+^2}, \tag{4.144}$$

$$B = \left(\frac{\rho^2 + (z - z_0)(z - Z_0) + r_- R_-}{\rho^2 + (z - z_0)(z + Z_0) + r_- R_+}\right)^{1/2} \frac{\rho^2 + z^2 - z_0^2 + r_- r_+}{2r_-^2}$$

so that for this new solution, the asymptotic behavior of the potentials f and A'_3 takes the form

$$f = 1 - \frac{2}{R}\left(Z_0 - \frac{4z_0\alpha^2}{1 + \alpha^2}\right) + \mathcal{O}\left(\frac{1}{R^3}\right),$$

$$A'_3 = \frac{4\alpha z_0}{(1 + \alpha^2)^2}\left[z_0(1 + 3\alpha^2) - Z_0(1 + \alpha^2)\right]\frac{\cos\vartheta}{R^2} + \mathcal{O}\left(\frac{1}{R^3}\right). \tag{4.145}$$

The total mass and the magnetic dipole moment in this case are then given by

$$m' = Z_0 - \frac{4z_0\alpha^2}{1 + \alpha^2},$$

$$\mu' = \frac{4\alpha z_0}{(1 + \alpha^2)^2}\left[z_0(1 + 3\alpha^2) - Z_0(1 + \alpha^2)\right]. \tag{4.146}$$

Transition to the Schwarzschild solution is achieved when $\mu'_0 = 0$ (at $\alpha = 0$). If $m' = 0$, we go over to the new solution for a massless magnetic dipole moment:

$$\mu_0 = 4\alpha z_0^2(1 - \alpha^2)(1 + \alpha^2)^{-2}. \tag{4.147}$$

In the case $z_0 = Z_0$, the above solution reduces to the solution (4.116).

Relations (4.133)–(4.135) permit the construction of other more complicated asymptotically flat solutions possessing the

Schwarzschild vacuum limit. For this purpose, e.g., instead of (4.136), one can take the function

$$\psi' = \psi + z_0\psi_0, \tag{4.148}$$

and instead of (4.145), the function

$$\psi' = \psi + \alpha\psi_0, \tag{4.149}$$

where ψ_0 is an arbitrary solution of the equation $\Delta\psi = 0$ which behaves as $\psi_0 = O(R^{-\alpha})$, $\alpha \geq 1$, when $R \to \infty$.

All the magnetostatic solutions considered in this section should be regarded as describing the Schwarzschild black hole in various magnetic fields, the potentials A_3 of which coincide at infinity with the classical expression for a magnetic dipole potential.

4.9 Method of Generating Static Einstein–Maxwell Fields

In this section, with the help of a method developed in Ref. [21], we present some particular solutions which reduce to the Schwarzschild metric in the case of vanishing magnetic (electric) field.

Using the results of Ref. [21] and equations (4.12), it is not difficult to show that equations (4.5) in the prolate ellipsoidal coordinates are satisfied by the functions

$$f = \frac{U^2}{4W^2}, \quad A_3'(\text{or } A_4) = \frac{V}{2W}, \tag{4.150}$$

where

$$U = (b_0c_0 - a_0d_0)(A^+B^- + A^-B^+),$$

$$V = (b_0c_0 + a_0d_0)(A^+B^- - A^-B^+) - 2a_0c_0B^-B^+ + 2b_0d_0A^-A^+,$$

$$W = c_0^2B^-B^+ - d_0^2A^-A^+ - c_0d_0(A^+B^- - A^-B^+),$$

where

$$A^{\pm} = (x+1)^{\delta}[(x^2-y^2)^{2\delta+2} - \alpha_0\beta_0(x+1)^2(x^2-1)^{2\delta+1}$$

$$\pm \beta_0(x+1)(y+1)(x^2-1)^{\delta}(x-y)^{2\delta+2}$$

$$\mp \alpha_0(x+1)(y-1)(x^2-1)^{\delta}(x+y)^{2\delta+2}],$$

$$B^{\pm} = (x-1)^{\delta}[(x^2-y^2)^{2\delta+2} - \alpha_0\beta_0(x-1)^2(x^2-1)^{2\delta+1}$$

$$\pm\,\beta_0(x-1)(y-1)(x^2-1)^{\delta}(x-y)^{2\delta+2}$$

$$\mp\,\alpha_0(x-1)(y+1)(x^2-1)^{\delta}(x+y)^{2\delta+2}]. \tag{4.151}$$

Functions (4.150) determine a class of asymptotically flat solutions of the static Einstein–Maxwell equations and contain seven arbitrary real constants $\delta, \alpha_0, \beta_0, a_0, b_0, c_0$ and d_0.

Let us consider the physically most interesting solutions arising from this series.

1. $\alpha_0 = \beta_0 = 0$, $a_0 = d_0, b_0 = c_0$, $\delta = \pm\frac{1}{2}$. In this case,

$$f = \frac{p_0^2(x^2-1)}{(p_0 x - 1)^2}, \quad A_4 = \frac{q_0}{p_0 x - 1}, \tag{4.152}$$

where

$$p_0 = \pm\frac{c_0^2 - d_0^2}{c_0^2 + d_0^2}, \quad q_0 = -\frac{2c_0 d_0}{c_0^2 + d_0^2}, \quad p_0^2 + q_0^2 = 1.$$

It is the Reissner–Nordstrom solution (4.22).

2. $\alpha_0 = -\beta_0$, $a_0 = d_0 = 0, b_0 = c_0$, $\delta = -\frac{1}{2}$. In this case, we have

$$f = \left(\frac{x-1}{x+1}\right)\frac{A^2}{C^2}, \quad A_3' = \frac{B}{C}, \tag{4.153}$$

where

$$A = [x^2 - y^2 + a_0^2(x^2-1)]^2 + 4\alpha_0 x^2(1-y^2),$$

$$B = 8\alpha_0^3 xy(x-1),$$

$$C = [x^2 - y^2 + a_0^2(x^2-1)]^2 - 4\alpha_0 y^2(x^2-1).$$

Integration of equations (4.4) in the prolate ellipsoidal coordinates gives

$$A_3 = \frac{4k_0\alpha_0^3(1-y^2)\left[2x^3(1+\alpha_0^2) + x^2(1-3\alpha_0^2) + y^2 + \alpha_0^2\right]}{(1+\alpha_0^2)\{[x^2-y^2+\alpha_0^2(x^2-1)]^2 + 4\alpha_0^2 x^2(1-y^2)\}}. \tag{4.154}$$

This solution was considered in Ref. [21].

3. $\alpha_0 = -\beta_0$, $a_0 = d_0, b_0 = c_0 = 0$, $\delta = \frac{1}{2}$. In this case, we have [22]

$$f = \left(\frac{x-1}{x+1}\right)\frac{A^2}{C^2}, \quad A_3' = \frac{B}{C}, \tag{4.155}$$

where

$$A = [(x^2 - y^2)^3 + \alpha_0^2(x^2 - 1)^3]^2 + 4\alpha_0^2 x^2(x^2 - 1)^2(1 - y^2)(x^2 + 3y^2),$$

$$B = 8\alpha_0 xy(x - 1)[(x^2 - y^2)^4 + 2\alpha_0^2(x^2 - 1)^3(x^2 + y^2)],$$

$$C = [(x^2 - y^2)^3 + \alpha_0^2(x^2 - 1)^2(x^2 + 1)]^2$$

$$- 4\alpha_0^2 y^2(x^2 - 1)\,(x + 1)^2(x^3 - 3x^2 + 3xy^2 - y^2).$$

4. $a_0 = d_0, b_0 = c_0 = 0$, $\delta = \frac{1}{2}$.

$$f = \left(\frac{x-1}{x+1}\right)\frac{A^2}{C^2}, \quad A_4 = \frac{B}{C}, \tag{4.156}$$

where

$$A = [(x^2 - y^2)^3 + \alpha_0\beta_0(x^2 - 1)^3]^2 + (x^2 - 1)^2(1 - y^2)$$

$$\times [\alpha_0(x + y)^3 - \beta_0(x - y)^3]^2,$$

$$B = 2(x - 1)\{(x^2 - y^2)^4[\alpha_0(x + y)^2 + \beta_0(x - y)^2]$$

$$- \alpha_0\beta_0(x^2 - 1)^3[\alpha_0(x + y)^4 + \beta_0(x - y)^4]\},$$

$$C = [(x^2 - y^2)^3 - \alpha_0\beta_0(x^2 - 1)^2(x + 1)^2]^2$$

$$- (x + 1)^2(x^2 - 1)[\alpha_0(y - 1)(x + y)^3 - \beta_0(y + 1)(x - y)^3]^2.$$

This solution was obtained in Ref. [27].

5. $\alpha_0 = 0$, $\beta_0 = \frac{q}{2p}$, $a_0 = d_0 = \sqrt{p - 1}, b_0 = c_0 = \sqrt{p + 1}$, $\delta = -\frac{1}{2}$, $p^2 - q^2 = 1$.

We obtain the expressions for f, A_3' and A_3.

In this case, we obtain the expressions for f, A_3' and A_3:

$$f = \frac{(x^2 - 1)\left[4p^2(x + y)^2 + q^2(1 - y^2)\right]^2}{D^2},$$

$$A_3' = -\frac{q^3(y^2 + 2xy + 1)}{D},$$

$$A_3 = -\frac{q^3 k_0 (1 - y^2)(2px + py + 1)}{p\left[4p^2(x + y)^2 + q^2(1 - y^2)\right]},$$

$$D = (x + p)\left[4p^2(x + y)^2 - q^2(x^2 - 1)\right]$$
$$+ q^2(x + y)\left[2(x^2 - 1) - (3p - 1)(x + y)\right]. \qquad (4.157)$$

This solution was considered in Ref. [24].

What can be said about the physical interpretation of the solutions (4.153), (4.155) and (4.157)? Apparently, they all should be considered as describing the Schwarzschild black hole in various magnetic fields, the potentials A_3 of which coincide at infinity with the classical expression for a magnetic dipole potential.

References

[1] H. Weyl, *Ann. Phys.* 54, 117 (1917).
[2] H. Reissner, *Ann. Phys.* 50, 106 (1916).
[3] G. Nordstrom, *Proc. Kon. Ned. Akad. Wet.* 20, 1238 (1918).
[4] S.D. Majumdar, *Phys. Rev.* 72, 390 (1947).
[5] A. Papapetrou, *Proc. R. Irish Acad.* A51, 191 (1947).
[6] R. Gautreau, R.B. Hoffman, *Nuovo Cim.* 16B, 162 (1973).
[7] W.B. Bonnor, *Proc. R. Soc. Lond.* A67, 225 (1954).
[8] M.A. Melvin, *Phys. Lett.* 8, 65 (1964).
[9] G.E. Tauber, *Canad. J. Phys.* 35, 477 (1957).
[10] W.B. Bonnor, Z. Phys. 190, 444 (1966).
[11] G. Daimois, *Memorial des sciences mathematique*, Fasc. XXV, Gauthier Vfflars, Paris (1927).
[12] R.M. Misra, D.B. Pandey, D.C. Srivastava, S.N. Tripathi, *Phys. Rev.* D7, 1582 (1973).
[13] M.Y. Wang, *Phys. Rev.* D9, 1835 (1974).
[14] J.R. Ray, M.S. Wei, *Nuovo Cim.* 42B, 151 (1977).
[15] M. Yamazaki, *Phys. Lett.* 67A, 337 (1978).
[16] M. Halisoy, *Gen. Rel. Grav.* 15, 1115 (1983).
[17] E. Herlt, *Gen. Rel Grav.* 9, 711 (1978).
[18] E. Herlt, *Gen. Rel. Grav.* 11, 337(1979).
[19] F.I. Cooperstock, V. Cruz, *Gen. Rel. Grav.* 10, 681 (1979).
[20] J. Carminati, F.I. Cooperstock, *Phys. Lett.* 94A, 27 (1983).
[21] T.I. Gutsunaev, V.S. Manko, *Gen. Rel. Grav.* 20, 327(1988).
[22] T.I. Gutsunaev, V.S. Manko, *Phys. Lett.* 132A, 85(1988).

[23] T.I. Gutsunaev, V.S. Manko, S.L. Elsgolts, *Class. Quant. Grav.* 6, 41 (1989).

[24] S.M. Abramyan, T.I. Gutsunaev, *Phys. Lett.* 144A, 437 (1990).

[25] T.I. Gutsunaev, S.L. Elsgolts, *Zh. Eksp. Teor. Fiz.* 104, 2257 (1993).

[26] T.I. Gutsunaev, D.A. Ponamarev, *Crav. Cosmol.* 3, 328 (1997).

[27] C. Hoenselaers, *Progr. Theor. Phys.* 67, 697 (1982).

[28] W.B. Bonnor, *Z. Phys.* 185, 439 (1961).

[29] D. Kramer, H. Stephani, E. Herlt, M. MacCallum, in E. Schmutzer (ed.), *Exact Solutions of Einstein's Field Equations*, Cambridge University Press, Cambridge (1980).

[30] D. Kramer, *Class. Quant. Grav.* 1, 45 (1984).

[31] R.S. Hanni, R. Ruffini, *Phys. Rev.* D8, 3259 (1973).

[32] T.I. Gutsunaev, A.A. Shaideman, *Gravit. Cosmol.* 8, 206 (2002).

[33] T.I. Gutsunaev, A.A. Shaideman, Z.E. Wael, *Gravit. Cosmol.* 13, 133 (2007).

Chapter 5

Stationary Solutions of the Einstein–Maxwell Fields

5.1 Introduction

In this chapter, we consider the Einstein–Maxwell equations in the case of a Stationary axially symmetric gravitational field outside the sources.

A stationary electrovacuum solution of the Einstein–Maxwell equations was offered in Ref. [1] in 1965. It pretends to describe the most common case of a charged rotating black hole. This solution can be obtained, for example, by applying Kramer–Neugebauer symmetry transformation [2] to Kerr's metric.

A method based on the theory of distributions, introduced by C. A. Lopez to analyze Kerr's singularity, was extended to the study of the Kerr–Newman geometry [3].

New methods of generating stationary Einstein–Maxwell fields were discovered in Refs. [4–14]. A class of approximate stationary solutions of the Einstein–Maxwell equations was found by Jamal N. Islam [15].

A physically meaningful problem is to seek a stationary solution of the Einstein–Maxwell equations reducing either to the Reissner–Nordstrém solution [16, 17] in the absence of the magnetic field or to the solution for a massive magnetic dipole, when the electric field vanishes. Fortunately, a possible solution [18] of this problem simply lies in applying the known symmetry transformation to the recently found solutions for a massive magnetic dipole [19], which result from

155

the superposition of the Bonnor solution [20] with an arbitrary static vacuum field.

5.2 Basic Equations

As was shown by Papapetrou [21], the line element describing the stationary axially symmetric gravitational field without the lack of generality can be represented on the following canonical form:

$$ds^2 = f^{-1}[e^{2\gamma}(d\rho^2 + dz^2) + \rho^2 d\varphi^2] - f(dt - \omega d\varphi)^2, \qquad (5.1)$$

where ρ, z, φ and t are respectively the canonical Weyl–Papapetrou coordinates and time; $f(\rho, z), \gamma(\rho, z), \omega(\rho, z)$ are three unknown functions which should be determined from the field equations.

The physical systems describing by the Papapetrou metric (5.1) in addition to the property of stationarity (characterized by the presence of the time-like Killing vector) possess one more symmetry: They are axially symmetrical.

The covariant metric tensor g_{ik}, in this case, has only six non-zero components:

$$
\begin{aligned}
g_{11} &= g_{22} = f^{-1}e^{2\gamma}, \\
g_{33} &= \rho^2 f^{-1} - f\omega^2, \\
g_{34} &= g_{43} = f\omega, \\
g_{44} &= -f,
\end{aligned}
\qquad (5.2)
$$

and the determinant of the matrix $//g_{ik}//$ is equal to

$$g = \det //g_{ik}// = -\left(\rho f^{-1}e^{2\gamma}\right)^2. \qquad (5.3)$$

For the signature of the spacetime $(+, +, +, -)$ which we use hereafter, the Einstein–Maxwell equations have the form [22, 23]

$$G_{ik} = 8\pi T_{ik}, \qquad (5.4)$$

$$\frac{\partial}{\partial x^k}\left(\sqrt{-g}F^{ik}\right) = 0, \qquad (5.5)$$

$$F_{ik,l} + F_{kl,i} + F_{li,k} = 0, \qquad (5.6)$$

where G_{ik} is the Einstein tensor relating to the Ricci tensor R_{ik} and to the curvature scalar R_{ik} by the relation

$$G_{ik} = R_{ik} - \frac{1}{2} g_{ik} R. \tag{5.7}$$

T_{ik} being the energy–momentum tensor of the electromagnetic field

$$T_{ik} = \frac{1}{4\pi} \left(F_{il} F_k^l - \frac{1}{4} F_{lm} F^{lm} g_{ik} \right), \tag{5.8}$$

where F_{ik} is the electromagnetic field tensor and comma denotes partial differentiation with respect to given coordinates.

To calculate the Ricci tensor and the curvature scalar, one should use the formulae

$$R_{ik} = \frac{\partial \Gamma_{ik}^l}{\partial x^l} - \frac{\partial \Gamma_{il}^l}{\partial x^k} + \Gamma_{ik}^l \Gamma_{lm}^m - \Gamma_{il}^m \Gamma_{km}^l, \tag{5.9}$$

$$R = R_{ik} g^{ik} \tag{5.10}$$

and the affine connections (Christoffel symbols) Γ_{kl}^i are found from the relations

$$\Gamma_{kl}^i = \frac{1}{2} g^{im} \left(\frac{\partial g_{mk}}{\partial x^l} + \frac{\partial g_{ml}}{\partial x^k} - \frac{\partial g_{kl}}{\partial x^m} \right). \tag{5.11}$$

The non-zero components of the symmetric Ricci tensor and the curvature scalar have the form

$$R_{11} = \rho^{-1} \gamma_{,\rho} - \frac{1}{4} f^{-2} \left[f_{,\rho}^2 - f_{,z}^2 - \rho^{-2} f^4 (\omega_{,\rho}^2 - \omega_{,z}^2) \right] + \frac{1}{2} f^{-2} F - \Gamma,$$

$$R_{12} = \rho^{-1} \gamma_{,z} - \frac{1}{2} f^{-2} (f_{,\rho} f_{,z} - \rho^{-2} f^4 \omega_{,\rho} \omega_{,z}),$$

$$R_{22} = -\rho^{-1} \gamma_{,\rho} + \frac{1}{4} f^{-2} \left[f_{,\rho}^2 - f_{,z}^2 - \rho^{-2} f^4 (\omega_{,\rho}^2 - \omega_{,z}^2) \right] + \frac{1}{2} f^{-2} F - \Gamma,$$

$$R_{33} = \frac{1}{2} e^{-2\gamma} \left[(\rho^2 f^{-2} + \omega^2) F + 2 f \omega W \right],$$

$$R_{34} = -\frac{1}{2} e^{-2\gamma} (f W + \omega F),$$

$$R_{44} = \frac{1}{2} e^{-2\gamma} F \tag{5.12}$$

and

$$R = 0. \tag{5.13}$$

Here, we introduced

$$F \equiv f\Delta f - (\vec{\nabla} f)^2 + \rho^{-2} f^4 (\vec{\nabla}\omega)^2,$$

$$W \equiv f(\omega_{,\rho,\rho} + \omega_{,z,z} - \rho^{-1}\omega_{,\rho}) + 2\vec{\nabla} f \vec{\nabla}\omega,$$

$$\Gamma \equiv \gamma_{,\rho,\rho} + \gamma_{,z,z} + \frac{1}{4} f^{-2} \left[(\vec{\nabla} f)^2 + \rho^{-2} f^4 (\vec{\nabla}\omega)^2 \right]. \tag{5.14}$$

The operators Δ and $\vec{\nabla}$ are defined by the formulae

$$\Delta \equiv \frac{\partial^2}{\partial\rho^2} + \rho^{-1}\frac{\partial}{\partial\rho} + \frac{\partial^2}{\partial z^2},$$

$$\vec{\nabla} \equiv \vec{\rho_0}\frac{\partial}{\partial\rho} + \vec{z_0}\frac{\partial}{\partial z} \tag{5.15}$$

($\vec{\rho_0}$ and $\vec{z_0}$ being unit vectors), i.e., they are similar to the ordinary Laplacian and gradient operators for the flat space written in the cylindrical coordinates provided that there is no dependence on the angular coordinate.

Calculating the components of the energy–momentum tensor T_{ik}, one should take into account that the antisymmetric electromagnetic field tensor F_{ik} can be chosen in the form

$$F_{ik} = A_{i,k} - A_{i,k} \tag{5.16}$$

that automatically fulfills the equations (5.6). The four-dimensional electromagnetic potential $A_i = (0, 0, A_3, -A_4)$ has in general case two components: the magnetic component A_3 and the electric A_4. Consequently, the non-zero components of the tensor F_{ik} have the form

$$F_{13} = -F_{31} = A_{3,\rho}, \quad F_{14} = -F_{41} = -A_{4,\rho},$$

$$F_{23} = -F_{32} = A_{3,z}, \quad F_{24} = -F_{42} = -A_{4,z}. \tag{5.17}$$

Then, calculations with the aid of (5.8) lead to the following expressions defining the components of the tensor T_{ik}:

$$T_{11} = \frac{1}{4\pi}\left[\rho^{-2}fA_{3,\rho}^2 + \left(\rho^{-2}f\omega^2 - f^{-1}\right)A_{4,\rho}^2 - 2\rho^{-2}f\omega A_{3,\rho}A_{4,\rho} - \frac{1}{2}A\right],$$

$$T_{12} = \frac{1}{4\pi}\left[\rho^{-2}f\left(A_{3,\rho} - \omega A_{4,\rho}\right)\left(A_{3,z} - \omega A_{4,z}\right) - f^{-1}A_{4,\rho}A_{4,z}\right],$$

$$T_{22} = \frac{1}{4\pi}\left[\rho^{-2}fA_{3,z}^2 + \left(\rho^{-2}f\omega^2 - f^{-1}\right)A_{4,z}^2 - 2\rho^{-2}f\omega A_{3,z}A_{4,z} - \frac{1}{2}A\right],$$

$$T_{33} = \frac{1}{4\pi}fe^{-2\gamma}\left[(\vec{\nabla}A_3)^2 - \frac{1}{2}\left(\rho^2 f^{-1} - f\omega^2\right)A\right],$$

$$T_{34} = -\frac{1}{4\pi}fe^{-2\gamma}\left[\vec{\nabla}A_3\vec{\nabla}A_4 + \frac{1}{2}f\omega A\right],$$

$$T_{44} = \frac{1}{4\pi}fe^{-2\gamma}\left[(\vec{\nabla}A_4)^2 + \frac{1}{2}fA\right], \tag{5.18}$$

where

$$A = \rho^{-2}f(\vec{\nabla}A_3)^2 + \left(\rho^{-2}f^2\omega^2 - f^{-1}\right)(\vec{\nabla}A_4)^2 - 2\rho^{-2}f\omega\vec{\nabla}A_3\vec{\nabla}A_4. \tag{5.19}$$

The components T_{13}, T_{14}, T_{23}, T_{24} and also, remembering the symmetry of the energy–momentum tensor, the components obtainable from these by the permutation of subscripts are equal to zero.

Substitution of the expressions (5.12), (5.18) in (5.4) leads to the following field equations:

$$\gamma_{,\rho} = \frac{1}{4}\rho f^{-2}[f^2{}_{,\rho} - f^2{}_{,z} - \rho^{-2}f^4(\omega^2{}_{,\rho} - \omega^2{}_{,z})] + \rho^{-1}f$$

$$\times [(A_{3,\rho} - \omega A_{4,\rho})^2 - (A_{3,z} - \omega A_{4,z})^2] - \rho f^{-1}(A_{4,\rho}^2 - A_{4,z}^2), \tag{5.20}$$

$$\gamma_{,z} = \frac{1}{2}\rho f^{-2}(f_{,\rho} f_{,z} - \rho^{-2}f^4\omega_{,\rho}\omega_{,z}) + 2\rho^{-1}f(A_{3,\rho} - \omega A_{4,\rho})$$

$$\times (A_{3,z} - \omega A_{4,z}) - 2\rho f^{-1}A_{4,\rho}A_{4,z}. \tag{5.21}$$

Let us consider now the Maxwell equations (5.5). Putting in them the value of the index i equal to 3 and then to 4, we obtain the equations of electromagnetism in the gravitational field.

When $i = 3$, we come to the equation

$$\rho^{-1}\left[\rho^{-1}f\left(A_{3,\rho} - \omega A_{4,\rho}\right)\right]_{,\rho} + \left[\rho^{-2}f\left(A_{3,z} - \omega A_{4,z}\right)\right]_{,z} = 0,$$

which may be written in the form

$$\vec{\nabla}\left[\rho^{-2}f\vec{\nabla}A_3 - \omega\vec{\nabla}A_4)\right] = 0, \tag{5.22}$$

if one introduces the differential operator similar to the axially symmetric operator of divergence of the vector $\vec{a}$ in the flat space defined in the cylindrical coordinates by

$$\vec{\nabla}\vec{a} = \rho^{-1}(\rho a_{,\rho})_{,\rho} + a_{,z,z}, \tag{5.23}$$

a_ρ and a_z being the projections of the vector $\vec{a}$ on the axes ρ and z, respectively.

When $i = 4$, we similarly obtain the equation

$$\vec{\nabla}\left[\rho^{-2}f^2\vec{\nabla}\omega - 4\rho^{-2}fA_4(\vec{\nabla}A_3 - \omega\vec{\nabla}A_4)\right] = 0. \tag{5.24}$$

Therefore, we have obtained all the Einstein–Maxwell equations in the case of the axially symmetric gravitational field outside the sources. We write them down together once again revealing the explicit form of F, W and Γ:

$$\vec{\nabla}\left[\rho^{-2}f(\vec{\nabla}A_3 - \omega\vec{\nabla}A_4)\right] = 0, \tag{I}$$

$$\vec{\nabla}\left[f^{-1}\vec{\nabla}A_4 + \rho^{-2}f\omega(\vec{\nabla}A_3 - \omega\vec{\nabla}A_4)\right] = 0, \tag{II}$$

$$\vec{\nabla}\left[\rho^{-2}f^2\vec{\nabla}\omega - 4\rho^{-2}fA_4(\vec{\nabla}A_3 - \omega\vec{\nabla}A_4)\right] = 0, \tag{III}$$

$$f\Delta f = (\vec{\nabla}f)^2 - \rho^{-2}f^4(\vec{\nabla}\omega)^2 + 2f(\vec{\nabla}A_4)^2 + 2\rho^{-2}f^3$$

$$\times (\vec{\nabla}A_3 - \omega\vec{\nabla}A_4)^2, \tag{IV}$$

$$\gamma_{,\rho} = \frac{1}{4}\rho f^{-2}\left[f_{,\rho}^2 - f_{,z}^2 - \rho^{-2}f^4(\omega_{,\rho}^2 - \omega_{,z}^2)\right] + \rho^{-1}f[(A_{3,\rho} - \omega A_{4,\rho})^2$$

$$- (A_{3,z} - \omega A_{4,z})^2] - \rho f^{-1}(A_{4,\rho}^2 - A_{4,z}^2), \tag{V}$$

$$\gamma_{,z} = \frac{1}{2}\rho f^{-2}\left[f_{,\rho}f_{,z} - \rho^{-2}f^4\omega_{,\rho}\omega_{,z}\right] + 2\rho^{-1}f(A_{3,\rho} - \omega A_{4,\rho})$$

$$\times (A_{3,z} - \omega A_{4,z}) - 2\rho f^{-1}A_{4,\rho}A_{4,z}, \tag{VI}$$

$$\gamma_{,\rho\,,\rho} + \gamma_{,z\,,z} = -\frac{1}{4}f^{-2}[(\vec{\nabla}f)^2 + \rho^2 f^4(\vec{\nabla}\omega)^2] + f^{-1}(\vec{\nabla}A_4)^2$$

$$+ \rho^{-2}f(\vec{\nabla}A_3 - \omega\vec{\nabla}A_4)^2. \tag{VII}$$

Of these seven field equations, the equations (I)–(VI) are independent, whereas (VII) is the consequence of the others.

Note also that the integrability condition of the equations (V), (VI) for the determination of the function γ is the system (I)–(IV) into which γ does not enter. Hence, the problem of obtaining exact axisymmetric solutions of the Einstein–Maxwell equations outside the sources reduces to the integration of a system of the differential equations (I)–(IV).

5.3 Ernst's Formulation of the Einstein–Maxwell Equations

Equations (I)–(IV) are a system of the four nonlinear differential equations in partial derivatives of the second order with respect to the unknown functions f, ω, A_3 and A_4 the integration of which is a difficult problem. Certain progress in finding the exact solutions of this system has been achieved after the publication of Ernst's paper [24] in which the equations (I)–(IV) were reduced to a pair of nonlinear differential equations for two complex functions. Consider now the procedure of obtaining the complex formulation of the Einstein–Maxwell equations mainly following [24].

Revealing in (I), the divergency operator one has

$$\left[\rho^{-1}f(A_{3,\rho} - \omega A_{4,\rho})\right]_{,\rho} + \left[\rho^{-1}f(A_{3,z} - \omega A_{4,z})\right]_{,z} = 0, \tag{5.25}$$

and it can be seen that (5.25) is the integrability condition for the existence of a new scalar potential A_3' defined by the formulae

$$A_{3,\rho}' = \rho^{-1}f(A_{3,z} - \omega A_{4,z}),$$
$$A_{3,z}' = -\rho^{-1}f(A_{3,\rho} - \omega A_{4,\rho}). \tag{5.26}$$

Adding the first relation of (5.26) multiplied by $\vec{z}_0$, to the second one multiplied by $\vec{\rho}_0$, we get as a result that equation (II) which may be written in the form

$$\vec{\nabla}\left(f^{-1}\vec{\nabla}A_4 + \rho^{-1}\omega\vec{\varphi}_0 \times \vec{\nabla}A_3'\right) = 0. \tag{5.27}$$

If one introduces a new complex potential

$$\Psi = A_4 + iA_3', \tag{5.28}$$

which satisfies the equation

$$\vec{\nabla}\left(f^{-1}\vec{\nabla}\Psi - i\rho^{-1}\omega\vec{\varphi}_0 \times \vec{\nabla}\Psi\right) = 0, \tag{5.29}$$

i being the imaginary unit.

In this case, the equation (III) now can be written in the form

$$\vec{\nabla}\left[\rho^{-2}f^2\vec{\nabla}\omega - 2\rho^{-1}\vec{\varphi}_0 \times \mathrm{Im}(\Psi^*\vec{\nabla}\Psi)\right] = 0, \tag{5.30}$$

where $\mathrm{Im}(t)$ denotes the imaginary part of t and $*$ denotes the complex conjugation.

Equation (5.30) may be regarded as the integrability condition for the existence of a new potential Φ, such that we come to the equation

$$\vec{\nabla}\left\{f^{-2}[\vec{\nabla}\Phi + 2\,\mathrm{Im}(\Psi^*\vec{\nabla}\Psi)]\right\} = 0. \tag{5.31}$$

On the other hand, the equation (IV) in terms of the new potentials Ψ, Φ takes the form

$$f\Delta f = (\vec{\nabla}f)^2 - \left[\vec{\nabla}\Phi + 2\,\mathrm{Im}(\Psi^*\vec{\nabla}\Psi)\right]^2 + 2f\vec{\nabla}\Psi\vec{\nabla}\Psi^* \tag{5.32}$$

and this can be combined with (5.31) if one introduces a complex function

$$\varepsilon = f - \Psi\Psi^* + i\Phi, \tag{5.33}$$

yielding as a result the equation

$$(\mathrm{Re}\varepsilon + \Psi\Psi^*)\,\Delta\varepsilon = (\vec{\nabla}\varepsilon + 2\Psi^*\vec{\nabla}\Psi^*)\vec{\nabla}\varepsilon, \tag{5.34}$$

where $\mathrm{Re}\,\varepsilon$ denotes the real part of the potential ε.

At the same time, equation (5.29) may be rewritten in the form

$$(\mathrm{Re}\,\varepsilon + \Psi\Psi^*)\,\Delta\Psi = \left(\vec{\nabla}\varepsilon + 2\Psi^*\vec{\nabla}\Psi\right)\vec{\nabla}\Psi, \tag{5.35}$$

which is not difficult to verify remembering that

$$f = \mathrm{Re}\,\varepsilon + \Psi\Psi^* \tag{5.36}$$

and also that from (5.30) follow the relations

$$\begin{aligned}
\omega_{,\rho} &= -\rho f^{-2}\,\mathrm{Im}(\varepsilon_{,z} + 2\Psi^*\Psi_{,z}), \\
\omega_{,z} &= \rho f^{-2}\,\mathrm{Im}(\varepsilon_{,\rho} + 2\Psi^*\Psi_{,\rho}).
\end{aligned} \tag{5.37}$$

Therefore, the system (I)–(IV) is reduced to a pair of nonlinear equations for the complex functions ε and Ψ.

The equations (V), (VI) for the metric coefficient γ can also be written in terms of the new functions ε and Ψ:

$$\begin{aligned}
\gamma_{,\rho} &= \frac{\rho}{4(\mathrm{Re}\,\varepsilon + \Psi\Psi^*)^2}\big[(\varepsilon_{,\rho} + 2\Psi^*\Psi_{,\rho})(\varepsilon^*_{,\rho} + 2\Psi\Psi^*_{,\rho}) \\
&\quad - (\varepsilon_{,z} + 2\Psi^*\Psi_{,z})(\varepsilon^*_{,\rho} + 2\Psi\Psi^*_{,\rho})\big] - \frac{\rho(\Psi_{,\rho}\Psi^*_{,\rho} - \Psi_{,z}\Psi^*_{,z})}{\mathrm{Re}\,\varepsilon + \Psi\Psi^*},
\end{aligned} \tag{5.38}$$

$$\gamma_{,z} = \frac{\rho\,\mathrm{Re}[(\varepsilon_{,\rho} + 2\Psi^*\Psi_{,\rho})(\varepsilon^*_{,z} + 2\Psi\Psi^*_{,z})]}{2(\mathrm{Re}\,\varepsilon + \Psi\Psi^*)^2} - \frac{2\rho\,\mathrm{Re}\left(\Psi^*_{,\rho}\Psi_{,z}\right)}{\mathrm{Re}\,\varepsilon + \Psi\Psi^*}.$$

Note that from any solution ε, Ψ one can construct a new solution ε' and Ψ' of the Einstein–Maxwell equations with the aid of the following symmetry transformations [25]:

$$\varepsilon' = \alpha\alpha^*\varepsilon, \qquad\qquad \Psi' = \alpha\Psi, \tag{5.39}$$

$$\varepsilon' = \varepsilon + ib, \qquad\qquad \Psi' = \Psi, \tag{5.40}$$

$$\varepsilon' = \varepsilon\,(1 + ic\varepsilon)^{-1}, \qquad \Psi' = (1 + ic\varepsilon)^{-1}, \tag{5.41}$$

$$\varepsilon' = \varepsilon - 2\beta^*\Psi - \beta\beta^*, \quad \Psi' = \Psi + \beta, \tag{5.42}$$

$$\begin{aligned}
\varepsilon' &= \varepsilon(1 - 2\gamma^*\Psi - \gamma\gamma^*\varepsilon)^{-1}, \\
\Psi' &= (\Psi + \gamma\varepsilon)(1 - 2\gamma^*\Psi - \gamma\gamma^*\varepsilon)^{-1},
\end{aligned} \tag{5.43}$$

where α, β, γ are complex and b, c are real constants.

In conclusion of this section, we give a more symmetric form of the system (5.34), (5.35).

Introducing new potentials ξ and η by the formulae

$$\varepsilon = (\xi - 1)(\xi + 1)^{-1}, \quad \Psi = \eta(\xi + 1)^{-1}, \tag{5.44}$$

one obtains the equations

$$(\xi\xi^* + \eta\eta^* - 1)\Delta\xi = 2(\xi^*\vec{\nabla}\xi + \eta^*\vec{\nabla}\eta)\vec{\nabla}\xi, \tag{5.45}$$

$$(\xi\xi^* + \eta\eta^* - 1)\Delta\eta = 2(\xi^*\vec{\nabla}\xi + \eta^*\vec{\nabla}\eta)\vec{\nabla}\eta, \tag{5.46}$$

into which ξ and η enter symmetrically.

5.4 Two Stationary Electrovacuum Generalizations of the Schwarzschild Solution

In this section, we present two stationary axisymmetric solutions of the Einstein–Maxwell equations which are asymptotically flat and have the Schwarzschild metric as a pure vacuum limit [18].

So, we first write down the line element representing a stationary axially symmetric field in its usual form

$$ds^2 = k^2 f^{-1}\left[e^{2\gamma}(x^2 - y^2)\left(\frac{dx^2}{x^2 - 1} + \frac{dy^2}{1 - y^2}\right)\right.$$

$$\left. + (x^2 - 1)(1 - y^2)d\varphi^2\right] - f(dt - \omega d\varphi)^2, \tag{5.47}$$

where three unknown functions f, γ and ω depend only on two prolate ellipsoidal coordinates x and y; then, Einstein–Maxwell equations are

$$
\begin{aligned}
(\mathrm{Re}\,\varepsilon + \Psi\Psi^*)\Delta\varepsilon &= (\vec{\nabla}\varepsilon + 2\Psi^*\vec{\nabla}\Psi)\vec{\nabla}\varepsilon, \\
(\mathrm{Re}\,\varepsilon + \Psi\Psi^*)\Delta\Psi &= (\vec{\nabla}\varepsilon + 2\Psi^*\vec{\nabla}\Psi)\vec{\nabla}\Psi,
\end{aligned}
\tag{5.48}
$$

where

$$
\begin{aligned}
\varepsilon = f - \Psi\Psi^* + i\Phi, \quad &\Psi \quad = A_4 + iA_3', \\
\mathrm{Re}\,\varepsilon = f - \Psi\Psi^*, \quad\quad &\mathrm{Im}\,\varepsilon = \Phi.
\end{aligned}
\tag{5.49}
$$

In equations (5.48) and (5.49), A_4 and A_3' are the electric and magnetic potentials, respectively, i is an imaginary unit, an asterisk denotes the complex conjugation and

$$
\Delta \equiv k^{-2}(x^2 - y^2)^{-1}\left\{ \frac{\partial}{\partial x}[(x^2 - 1)\frac{\partial}{\partial x}] + \frac{\partial}{\partial y}\left[(1 - y^2)\frac{\partial}{\partial y}\right] \right\},
\tag{5.50}
$$

$$
\vec{\nabla} \equiv k^{-1}(x^2 - y^2)^{-1/2}\left(\vec{x}_0\sqrt{x^2 - 1}\frac{\partial}{\partial x} + \vec{y}_0\sqrt{1 - y^2}\frac{\partial}{\partial y} \right)
\tag{5.51}
$$

($\vec{x}_0$ and $\vec{y}_0$ are unit vectors), thus ω is connected with ε and Ψ by the relations (5.37). If ε and Ψ are known, the metric function γ can be found from equation (5.38).

We used the symmetry transformation presented in Ref. [2] to construct the solution of the form

$$
\varepsilon = \frac{\varepsilon_0 + 2\beta\Psi_0 - \beta^2}{1 - 2\beta\Psi_0 - \beta^2\varepsilon_0}, \quad \Psi = \frac{(1 + \beta^2)\Psi_0 + \beta(\varepsilon_0 - 1)}{1 - 2\beta\Psi_0 - \beta^2\varepsilon_0},
\tag{5.52}
$$

where β is a real constant, whereas ε_0 and Ψ_0 are either of the two "seed" magnetostatic solutions (4.116)

$$
\varepsilon_0^{(1)} = \frac{x - 1}{x + 1} \cdot \frac{[x^2 - y^2 + \alpha^2(x + 1)^2]^2 - 4\alpha^2 y^2(x^2 - 1)}{[x^2 - y^2 + \alpha^2(x - 1)^2]^2 - 4\alpha^2 y^2(x^2 - 1)},
\tag{5.53}
$$

$$
\Psi_0^{(1)} = \frac{8i\alpha^3 xy(x - 1)}{[x^2 - y^2 + \alpha^2(x - 1)^2]^2 - 4\alpha^2 y^2(x^2 - 1)}
$$

or (4.125)

$$\varepsilon_0^{(2)} = \frac{x-1}{x+1} \cdot \frac{B_-}{B_+}, \qquad \Psi_0^{(2)} = -\frac{8i\alpha y C}{B_+},$$

$$B_\pm = [(x^2 - y^2)^3 + \alpha^2(x \pm 1)^2(x^2-1)^2]^2$$

$$- 4\alpha^2 y^2(x \pm 1)^2(x^2-1)[(x^3 \mp 3x^2 + 3xy^2 \mp y^2)^2],$$

$$C = (x-1)[(x^2-y^2)^4 + 2\alpha^2(x^2+y^2)(x^2-1)^3]. \tag{5.54}$$

Here, α is a real constant, and the prolate ellipsoidal coordinates (x, y) are related to ρ and z by

$$x = \frac{r_+ + r_-}{2k}, \qquad y = \frac{r_+ - r_-}{2k},$$

$$r_\pm = \left[\rho^2 + (z \pm k)^2\right]^{1/2}, \qquad k = \text{const.} \tag{5.55}$$

The new metric coefficient f is given by the expression

$$f = \frac{(1-\beta^2)^2(\varepsilon_0 - \Psi_0^2)}{(1-\beta^2\varepsilon_0)^2 - 4\beta^2\Psi_0^2} \tag{5.56}$$

and the functions γ and ω are given by

$$\exp(2\gamma^{(1)}) = \frac{x^2-1}{x^2-y^2} \cdot \frac{\langle[x^2-y^2+\alpha^2(x^2-1)]^2 + 4\alpha^2 x^2(1-y^2)\rangle^4}{(1+\alpha^2)^8(x^2-y^2)^8},$$

$$\omega^{(1)} = \frac{16k\alpha^3\beta(1-y^2)}{(1+\alpha^2)(1-\beta^2)^2} \tag{5.57}$$

$$\cdot \frac{2(1-\beta^2)(1+\alpha^2)x^3 + (1+\beta^2)[(1-3\alpha^2)x^2 + y^2 + \alpha^2]}{[x^2-y^2+\alpha^2(x^2-1)]^2 + 4\alpha^2 x^2(1-y^2)}$$

for the first solution and

$$\exp(2\gamma^{(2)}) = \frac{x^2-1}{x^2-y^2} \cdot \frac{D^4}{(1+\alpha^2)^8(x^2-y^2)^{24}},$$

$$\omega^{(2)} = -\frac{16\alpha\beta k(1-y^2)E}{(1+\alpha^2)D},$$

$$D \equiv [(x^2 - y^2)^3 + \alpha^2(x^2 - 1)^3]^2 + 4\alpha^2 x^2(1 - y^2)(x^2 - 1)^2$$
$$\times (x^2 + 3y^2)^2,$$

$$E \equiv (x^2 - y^2)^4[2(1 - \beta^2)x^3 + (1 + \beta^2)(x^2 + y^2)]$$
$$- \alpha^2 \langle[(1 - \beta^2)(x^2 + 1) + 2(1 + \beta^2)x]$$
$$\times [-6x^9 + 4x^7 y^2 - 2x^5(6y^4 + 21y^2 + 1)$$
$$+ 4x^3 y^2(2y^4 - 17y^2 - 2) - 2xy^4(y^4 + y^2 - 1)]$$
$$+ [(1 + \beta^2)(x^2 + 1) + 2(1 - \beta^2)x]$$
$$\times [5x^8 + x^6(13y^2 + 3) + x^4(40y^4 + 43y^2 - 1)$$
$$- 3x^2 y^2(y^4 - 11y^2 + 2) + y^4(y^4 + y^2 - 1)]\rangle$$
$$+ \alpha^4(x^2 - 1)^2 \langle 4x(x^2 + y^2 + 1)[(1 - \beta^2)(x^4 + 6x^2 + 1)$$
$$+ 4(1 + \beta^2)x(x^2 + 1)] - (6x^2 + y^2 + 1)$$
$$\times [(1 + \beta^2)(x^4 + 6x^2 + 1) + 4(1 - \beta^2)x(x^2 + 1)]\rangle$$

$$\tag{5.58}$$

for the second solution.

The total mass M, angular momentum J, charge Q and magnetic dipole moment μ for each of the solutions are then found to be

$$M^{(1)} = \frac{(1 + \beta^2)k(1 - 3\alpha^2)}{(1 - \beta^2)(1 + \alpha^2)},$$

$$J^{(1)} = \frac{16\alpha^3 \beta k^2}{(1 - \beta^2)(1 + \alpha^2)^2},$$

$$Q^{(1)} = \frac{-2\beta k(1 - 3\alpha^2)}{(1 - \beta^2)(1 + \alpha^2)},$$

$$\mu^{(1)} = \frac{8\alpha^3(1 + \beta^2)k^2}{(1 - \beta^2)(1 + \alpha^2)^2},$$

$$\tag{5.59}$$

for the first solution, and

$$M^{(2)} = \frac{(1+\beta^2)k(1+5\alpha^2)}{(1-\beta^2)(1+\alpha^2)},$$

$$Q^{(2)} = \frac{-2\beta k(1+5\alpha^2)}{(1-\beta^2)(1+\alpha^2)},$$

$$J^{(2)} = -\frac{16\alpha\beta(1+2\alpha^2)k^2}{(1-\beta^2)(1+\alpha^2)^2},$$

$$\mu^{(2)} = -\frac{8\alpha(1+\beta^2)(1+2\alpha^2)k^2}{(1-\beta^2)(1+\alpha^2)^2}$$

$$(5.60)$$

for the second one.

The case $\beta = 0$ (i.e., $Q = 0$) leads to the magnetostatic solutions for the massive magnetic dipole [19].

When $\alpha = 0$ (i.e, $\mu = 0$), we get the Reissner–Nordstrém solution.

Lastly, when the electromagnetic field is equal to zero ($\alpha = \beta = 0$), we come to the Schwarzschild metric.

References

[1]   E.T. Newman, E. Couch, K. Chinnapared, A. Exton, A. Prakash, R. Torrence, *J. Math. Phys.* 6(6), 918 (1965).
[2]   D. Kramer, G. Neugebauer, *Ann. Phys.* 24, 59 (1969).
[3]   C.A. Lopez, *Nuovo Cim.* B76(1), 9 (1983).
[4]   R. Geroch, *J. Math. Phys.* 12, 918 (1971).
[5]   W. Kinnersley, *J. Math. Phys.* 14, 651 (1973).
[6]   W. Israel, G.A. Wilson, *J. Math. Phys.* 13, 865 (1972).
[7]   M. Demianski, *Acta Phys. Pol.* B7(8), 567 (1976).
[8]   J.R. Ray, M.S. Wei, *Nuovo Cim.* B42(1), 151 (1977).
[9]   F.I. Cooperstock, V. Cruz, *Gen. Rel. Grav.* 10(8), 681 (1979).
[10] P. Wils, N. Van den Bergh, *Class. Quant. Grav.* 1, 193 (1984).
[11] M. Halilsoy, *Lett. Nuovo Cim.* 37(6), 231 (1983).
[12] B. Leaute, G. Marcilhacy, *Lett. Nuovo Cim.* 40(4), 102 (1984).
[13] A. Diaz, *J. Math. Phys.* 26(1), 155, 141f (1985).
[14] A. Chamorro, V.S. Manko, T.E. Denisova, *Phys. Rey. D*, 44(10), 3147 (1991).
[15] J.N. Islam, *Gen. Rel. Grav.* 7(8), 669 (1976).
[16] H. Reissner, *Ann. Phys.* 50, 106 (1916).

[17] G. Nordstrém, *Proc. Kon. Ned. Akad. Wet.* 20, 1238 (1918).

[18] T.I. Gutsunaev, V.S. Manko, *Phys. Rev. D*, 40(6), 2141 (1989).

[19] T.I. Gutsunaev, V.S. Manko, *Phys. Lett.* A132, 85 (1988).

[20] W.B. Bonnor, *Z. Phys.* 190, 444 (1966).

[21] A. Papapetrou, *Ann. Phys.* 12, 309 (1953).

[22] L.D. Landau, E.M. Lifshitz, *The Classical Theory of Fields*, Pergamon Press, Oxford (1983).

[23] C.W. Misner, K.S. Thorne, J.A. Wheeler, *Gravitation*, Freeman, San Francisco (1973).

[24] F.J. Emst, *Phys. Rev.* 168, 1415 (1968).

[25] D. Kramer, H. Stephani, M.A.H. MacCallum, *Exact Solutions of Einstein's Field Equations*, VEB Deutscher Verlag der Wissenschaften, Berlin (1980).